MICROWAVE TRANSMISSION LINE COUPLERS

The Artech House Microwave Library

Adamy, David L., **Preparing and Delivering Effective Technical Presentations**

Algeri, Salvatore, W. Stephen Cheung, and Lawrence A. Stark, **Microwaves Made Simple — the Workbook**

Arbenz, Kurt, and Alfred Wohlhauser, **Advanced Mathematics for Practicing Engineers**

Cheung, W. Stephen, Frederick H. Levien, et al., **Microwaves Made Simple: Principles and Applications**

Craven, George F., and Richard F. Skedd, **Evanescent Mode Microwave Components**

DiLorenzo, James V., and Deen D. Khandelwal, **GaAs FET Principles and Technology**

Engelson, Morris, **Modern Spectrum Analyzer Theory and Applications**

Erst, Stephen J., **Receiving Systems Design**

Gardiol, Fred E., **Introduction to Microwaves**

Gardiol, Fred E., **Lossy Transmission Lines**

Garver, Robert V., **Microwave Diode Control Devices**

Gilmour, A.S., Jr., **Microwave Tubes**

Granatstein, Victor L., and Igor Alexeff, eds., **High-Power Microwave Sources**

Laverghetta, Thomas S., **Microwave Measurements and Techniques**

Laverghetta, Thomas S., **Handbook of Microwave Testing**

Laverghetta, Thomas S., **Solid-State Microwave Devices**

Lurie, B.J., **Feedback Maximization**

Robert, Philippe, **Electrical and Magnetic Properties of Materials**

Saad, Theodore, ed., **Microwave Engineer's Handbook,** 2 Vols.

Smith, W.V., **Laser Applications**

Ulaby, F.T., R.K. Moore, and A.K. Fung, **Microwave Remote Sensing,** 3 Vols.

Whalen, Timothy, **Writing and Managing Winning Technical Proposals**

Williams, Ralph E., **Gallium Arsenide Processing Techniques**

MICROWAVE TRANSMISSION LINE COUPLERS

J.A.G. MALHERBE

Artech House

Library of Congress Cataloging-in-Publication Data

Malherbe, J.A.G., 1940—
Microwave transmission line couplers.

Bibliography: p.
Includes index.
1. Directional couplers. I. Title
TK7871.65.M35 1988 621.381'331 88-19338
ISBN 0-89006-300-1

ARTECH HOUSE, INC.
685 Canton Street
Norwood, MA 02062

International Standard Book Number: 0-89006-300-1
Library of Congress Catalog Card Number: 88-19338

10 9 8 7 6 5 4 3 2 1

To

Louis

Contents

Preface

This book deals with directional couplers constructed of TEM-mode transmission lines, excluding dispersive waveguides. Its aims are twofold. First, we develop a detailed and thorough theoretical basis that is intended not only to further the understanding of the mechanisms of the various networks, but also to understand the underlying analytical approach. We think that it is both necessary and convenient to have all the background material available in one place.

The second aim is to provide a reference for the engineer interested in the design of a directional coupler. Thus, the text also covers more practical considerations by including a large number of complete design examples; here, the accent is on the design procedure itself, and two simple computer programs for the analysis of symmetrical three- and four-port networks are included. The examples in the text are all analyzed by using these programs. A simple program for the design of physical structures is also included.

The book is complete to a certain extent, in that it briefly summarizes the most relevant results from network theory in Chapter 1. The properties and performance of single-section connected-line couplers and power dividers are treated in Chapter 2. Simple computer analysis of the even and odd networks show the limited bandwidth, and the design and properties of multisection couplers is treated. Power dividers are treated similarly through synthesis and analysis.

Various coupled line cross sections are treated in Chapter 3, including all the relevant types of coupled striplines and microstriplines. Chapter 4 describes the synthesis of single-section coupled line couplers, illustrating the limitations in coupling of various forms of physical construction. Solutions to the problem of tight coupling such as the Lange coupler are described.

Chapter 5 treats the basic principle of a multisection coupler, and derives the conditions under which a multisection coupler can function. The various types of stepped couplers are studied. Directional couplers making use of continuously varying coupling are treated in Chapter 6. The conditions for operation are investigated and design based on the Hecken taper are treated.

The properties of coupled lines are basic to the synthesis and design of coupled-line couplers. Appendix B covers the most important coupled-line configurations from an applications point of view; that is, no attention is given to the derivation of design formulas, *et cetera,* but rather to the application of coupled lines to coupled-line coupler design, and their calculation by means of the computer program, LDIMS.

The author acknowledges with gratitude the general assistance of Julia Bekker and Elsabé Malherbe who helped with the typing of some of the tables, and Jaco Coetzee who plotted the figures.

J. A. G. MALHERBE
PRETORIA
MARCH 1988

Chapter 1
Basic Principles and Definitions

1.1 INTRODUCTION

Directional couplers find wide application in microwave networks and systems. They perform a variety of functions, such as splitting and combining power in mixers (hybrids), sampling power from sources for level control, separating incident and reflected signals in network analyzers, or dividing power among a number of loads. In this book, four types of networks will be studied that, in the general definition, can all be considered to be directional couplers: coupled-line couplers and power dividers, and connected-line couplers and power dividers. The differences will be clarified later.

As implied in the title, all the networks will be transmission-line or TEM-networks, stripline or microstrip line, even though the latter is not strictly a TEM structure. Waveguide couplers, the class of couplers making use of the principle of forward coupling, will not be included.

All directional couplers have the special property of coupling energy from a specific direction into a coupled port; there is usually an isolated port to which the coupled power from the driven port is extremely small, thus the name directional. In the case of power dividers, there may be a number of coupled and mutually isolated ports, and the power division is usually in phase. The function of the ports generally cannot be rotated.

The coupled-line couplers make use of the backward wave coupling principle; the odd and even modes of the wave on the pair of coupled lines travel at the same velocity but, due to the different characteristic impedances, cancel at the isolated port and combine in the correct way at the coupled and through ports. These couplers and power dividers can be designed to have extremely wide bandwidths but become complex when tight coupling has to be achieved at the same time.

The connected-line couplers include all couplers where the mechanism is achieved by an ohmic connection rather than by means of coupled modes. This includes branch-line couplers, rat-race or ring couplers, and power dividers such as the Wilkinson power divider.

1.2 NETWORK PARAMETERS

The properties and performance of most networks in the microwave frequency range are described in terms of the scattering parameters, and it is necessary to briefly define the various parameters and their network and matrix properties. As this book is not primarily concerned with aspects of network theory, only the relevant material will be repeated and the reader will be referred to the literature where necessary.

1.2.1 Scattering Parameters

The scattering (S) parameters relate the normalized incident and scattered waves at any given port in a network. If all ports are terminated in a characteristic impedance Z_0, then the waves can be considered as either voltage or current waves, and the S-parameters become the reflection and transmission coefficients at and between the respective ports. Because these parameters describe a network, they can be shown to be related [1] to the element parameters. If $[y]$ and $[z]$ are the admittance and impedance matrices of a network, normalized to Z_0, then,

$$[z] = (I + S)(I - S)^{-1} \tag{1.1}$$

$$[y] = (I - S)(I + S)^{-1} \tag{1.2}$$

$$[S] = (z - I)(z + I)^{-1} \tag{1.3}$$

$$[S] = (I - y)(I + y)^{-1} \tag{1.4}$$

where I denotes the unity matrix. Thus, the network performance properties, S, can be readily converted to element specifications by means of eqs. (1.1) and (1.2).

The scattering matrix has a very important property for lossless networks, the so-called unitary property, stated as

$$[S]^*[S] = [I] \tag{1.5}$$

where $[S]^*$ denotes the complex conjugate of $[S]$. Directional couplers can normally be described by a 4×4 matrix, so that if the scattering matrix is given by

$$[S] = \begin{bmatrix} S_{11} & S_{12} & S_{13} & S_{14} \\ S_{21} & S_{22} & S_{23} & S_{24} \\ S_{31} & S_{32} & S_{33} & S_{34} \\ S_{41} & S_{42} & S_{43} & S_{44} \end{bmatrix} \tag{1.6}$$

then,

$$\sum_{j=1}^{4} |S_{ij}|^2 = 1, \qquad i = 1 \rightarrow 4 \tag{1.7}$$

or, for example, for $i = 1$,

$$|S_{11}|^2 + |S_{12}|^2 + |S_{13}|^2 + |S_{14}|^2 = 1 \tag{1.8}$$

Additionally,

$$\sum_{i=1}^{4} S_{jm} \cdot S_{jn}^* = 0, \qquad m \neq n \tag{1.9}$$

As an example, for columns 2 and 4,

$$S_{12} \cdot S_{14}^* + S_{22} \cdot S_{24}^* + S_{32} \cdot S_{34}^* + S_{42} \cdot S_{44}^* = 0 \tag{1.10}$$

Application of these principles will be seen to be a very powerful tool in the analysis of directional couplers, as it enables us to make specific statements about the performance of circuits and to predict properties that are not known *a priori.*

1.2.2 Odd and Even Networks

In contrast to the scattering-parameter approach, which is a synthesis tool *par excellence,* the use of odd and even networks is ideally suited to the analysis of known circuit configurations. Especially in more complex cases, forming an equivalent circuit by means of the odd and even networks and then analyzing or synthesizing the equivalent circuit is an extremely powerful analytical method.

The description of most symmetrical networks can be greatly simplified by using the known properties of symmetry. This means that the analyses of four-port networks simplify to the description of two two-port networks, called the odd and even networks. The two ports on either side of the line of symmetry are driven by in-phase generators, and due to the equal potentials of the two sides of the network, a magnetic conductor (electric open circuit) can be introduced along the plane of symmetry, as shown in Figure 1.1(a).

With the two generators driven out of phase, the plane of symmetry of the network lies at zero potential and can consequently be grounded. Figures 1.1(b) and 1.1(c) show the resultant even- and odd-mode networks. There is a simple relationship between the scattering coefficients on either side of the line of symmetry of a symmetrical network and the odd- and even-mode input admittances [2].

If port j is symmetrically situated with respect to port i,

$$S_{ij} = \frac{y_e - y_o}{(1 + y_e)(1 + y_o)}. \tag{1.11}$$

Note that this type of analysis is not limited to application at a fixed frequency.

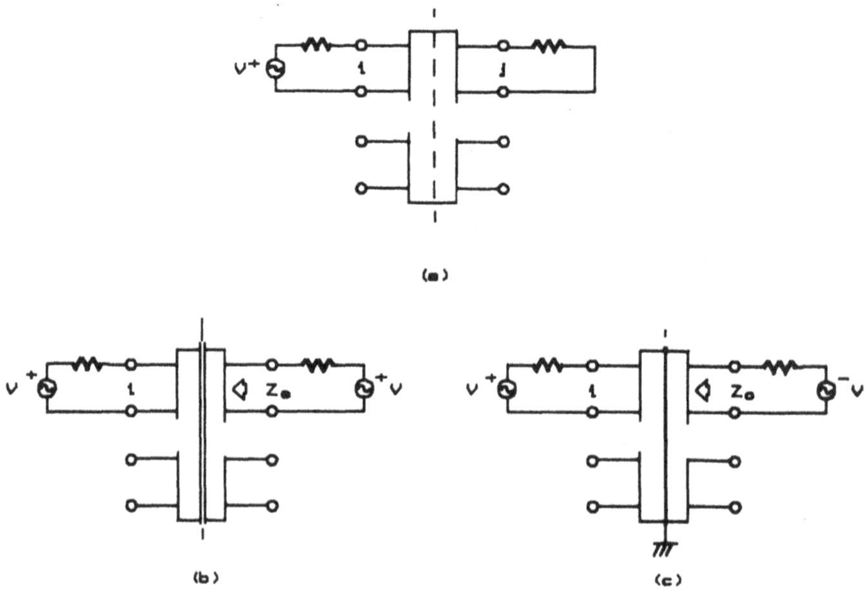

Figure 1.1 A general symmetrical network is shown in (a), and fed as even and odd networks in (b) and (c).

The even- and odd-mode admittances y_e and y_o are also related to the reflection coefficient of a symmetrical network as,

$$S_{11} = \frac{1 - y_e y_o}{(1 + y_e)(1 + y_o)} \tag{1.12}$$

where the even- and odd-mode admittances refer to the specific port under consideration.

1.2.3 Transmission Line Properties

Of prime importance in the study of any transmission line problem are the familiar transmission line equations. The input impedance to a section of transmission line of electrical length βl and of characteristic impedance Z_0, terminated in a load of Z_L is given by,

$$Z_{\text{in}} = Z_0 \frac{Z_L + jZ_0 \tan\beta l}{Z_0 + jZ_L \tan\beta l} \tag{1.13}$$

while the voltage at the generator, V_G, is related to the voltage at the load by

$$V_G = V_L(\cos\beta l + jZ_0/Z_L \sin\beta l) \tag{1.14}$$

The input impedance to a short-circuited section of line is given by

$$Z_{\text{in}}(0) = jZ_0 \tan\beta l \tag{1.15}$$

while for the open-circuited line,

$$\begin{aligned} Z_{\text{in}}(\infty) &= -jZ_0 \cot\beta l \\ Y_{\text{in}}(0) &= jY_0 \tan\beta l \end{aligned} \tag{1.16}$$

From time to time these two quantities will be referred to as Richards' inductors and capacitors respectively, because, under Richards' transform [3], [2]

$$\Omega \equiv \tan\beta l = \tan\theta \tag{1.17}$$

$$\begin{aligned} Z_{\text{in}}(0) &= j\Omega Z_0 = j\Omega L \\ Y_{\text{in}}(\infty) &= j\Omega Y_0 = j\Omega C \end{aligned} \tag{1.18}$$

1.2.4 Transmission Line Network Model

In this book network principles will be used where possible in order to simplify the complex electromagnetic problems. In describing coupled lines, the most general way of arriving at network properties is to make use of the network model described by Malherbe [2]. Figure 1.2(a) shows a schematic representation of a pair of coupled lines, the exact cross section of which need not be known (at this stage), provided that they support a TEM-mode of propagation. The lines can also be described in the equivalent circuit shown in Figure 1.2(b), in which the line parameters are given by

$$C_{11}/\epsilon = \frac{\eta}{\sqrt{\epsilon_r}}\left[\frac{1}{Z_1} - \frac{n-1}{Z_2}\right] \tag{1.19}$$

$$C_{12}/\epsilon = \frac{\eta}{\sqrt{\epsilon_r}} \cdot \frac{n}{Z_2} \tag{1.20}$$

$$C_{22}/\epsilon = \frac{\eta}{\sqrt{\epsilon_r}} \cdot \frac{n(n-1)}{Z_2} \tag{1.21}$$

The capacitances in eqs. (1.19)–(1.21) are related to the physical capacitance values through

$$C_{11} = 2C_{p1} + 2C_f + 2C_{fe} \tag{1.22}$$

$$C_{22} = 2C_{p2} + 2C_f + 2C_{fe} \tag{1.23}$$

$$C_{12} = C_{fo} - C_{fe} = \Delta C \tag{1.24}$$

These values are further connected to the conventional odd- and even-mode impedances and capacitances of coupled lines through the following relationships.

In general,

$$Z = \frac{\eta}{\sqrt{\epsilon_r}} \cdot \frac{1}{C/\epsilon} \tag{1.25}$$

$$n = 1 + \frac{C_{22}}{C_{12}} \tag{1.26}$$

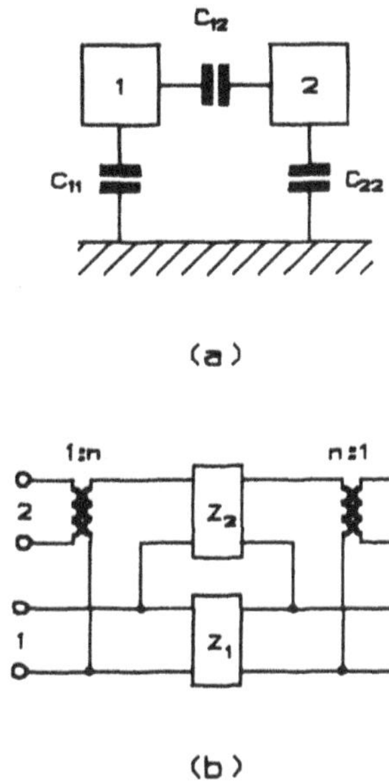

Figure 1.2 **Capacitances of pair of coupled lines (a), with the equivalent network model from [2] shown in (b).**

For lines of equal width, $C_{11} = C_{22}$, and from eqs. (1.19) and (1.21),

$$Z_2 = Z_1(n^2 - 1) \tag{1.27}$$

so that, in terms of the even- and odd-mode impedances,

$$Z_{0e} = \frac{\eta}{\sqrt{\epsilon_r}} \cdot \frac{1}{C_{11}/\epsilon} \tag{1.28}$$

$$Z_{0o} = \frac{\eta}{\sqrt{\epsilon_r}} \cdot \frac{1}{C_{12}(n + 1)} \tag{1.29}$$

$$Z_1 = \tfrac{1}{2}(Z_{0e} + Z_{0o}) \tag{1.30}$$

Use of the network model follows a number of straightforward rules. The relevant terminating conditions are applied to the ports, and the network is then simplified. Open-circuit sections of transmission line become capacitors, and short-circuited sections become inductors, in accordance with eq. (1.18).

One rule that seems trivial but is in fact extremely important in the simplification of circuits is that the current must always be the same on the two legs of a transmission line at the same set of terminals. Thus, if one leg of a line is open-circuited, the other may be disconnected from any part of the circuit without affecting the rest of the circuit.

1.2.5 *ABCD* Parameters

While scattering parameters are convenient for the description of microwave performance, the *ABCD* (also called chain, cascade, general-circuit) parameters are well suited to the description of cascade networks. We can define the properties of ladder-type networks in terms of their cascade matrices, making it possible to perform cascade synthesis of the network properties. The convenience arises because the cascade matrix of a number of cascade sections is simply the product of their cascade matrices.

The cascade matrix is related to the other matrices [1] as follows. For the Z-matrix:

$$\begin{bmatrix} Z_{11} = \dfrac{A}{C} & Z_{22} = \dfrac{D}{C} \\[2ex] Z_{12} = \dfrac{AD}{C} - B & Z_{21} = \dfrac{1}{B} \end{bmatrix} \tag{1.31}$$

$$\begin{bmatrix} A = \dfrac{Z_{11}}{Z_{21}} & C = \dfrac{1}{Z_{21}} \\[2ex] B = \dfrac{Z_{11}Z_{22} - Z_{12}Z_{21}}{Z_{21}} & D = \dfrac{Z_{22}}{Z_{21}} \end{bmatrix} \tag{1.32}$$

for the Y-matrix:

$$\begin{bmatrix} Y_{11} = \dfrac{D}{B} & Y_{22} = \dfrac{A}{B} \\[2ex] Y_{12} = -\dfrac{1}{B} & Y_{21} = C - \dfrac{AD}{B} \end{bmatrix} \tag{1.33}$$

and for the S-matrix:

$$\begin{bmatrix} S_{11} = \dfrac{A + B - C - D}{A + B + C + D} & S_{22} = \dfrac{-A + B - C + D}{A + B + C + D} \\[2ex] S_{12} = \dfrac{2(AD - BC)}{A + B + C + D} & S_{21} = \dfrac{2}{A + B + C + D} \end{bmatrix} \tag{1.34}$$

The cascade matrix has certain other properties. If the network is reciprocal:

$$AD - BC = 1 \tag{1.35}$$

If the network is symmetrical:

$$A = D \tag{1.36}$$

In terms of the scattering parameters, reciprocity gives

$$AD - BC = \frac{S_{12}}{S_{21}} = 1 \tag{1.37}$$

$$S_{12} = S_{21} = \frac{2}{A + B + C + D} \tag{1.38}$$

while, from symmetry:

$$S_{11} = S_{22} = \frac{B - C}{2A + B + C} \tag{1.39}$$

Various elements can be described by the $[A]$ matrix. The following results are useful.

The cascade matrix of a series connected impedance, Z:

$$\begin{bmatrix} A & B \\ C & D \end{bmatrix} = [A] = \begin{bmatrix} 1 & Z \\ 0 & 1 \end{bmatrix} \tag{1.40}$$

and for a shunt admittance, Y:

$$[A] = \begin{bmatrix} 1 & 0 \\ Y & 1 \end{bmatrix} \tag{1.41}$$

while, for a unit element of impedance, Z_0, and electrical length βl, we have

$$\begin{aligned} [A] &= \begin{bmatrix} \cos\beta l & jZ_0 \sin\beta l \\ jY_0 \sin\beta l & \cos\beta l \end{bmatrix} \\ &= \frac{j}{\sqrt{\tan^2\beta l - 1}} \begin{bmatrix} 1 & jZ_0 \tan\beta l \\ jY_0 \tan\beta l & .\ 1 \end{bmatrix} \\ &= \frac{1}{\sqrt{1 - S^2}} \begin{bmatrix} 1 & Z_0 S \\ S/Z_0 & 1 \end{bmatrix} \end{aligned} \tag{1.42}$$

where the complex Richards' frequency is given by

$$S = j\Omega = j \tan\beta l \tag{1.43}$$

Finally, the phase shift through a network is given in terms of the cascade parameters by

$$\cos\varphi = \tfrac{1}{2}(A + D) \tag{1.44}$$

1.2.6 Interrelationship

The various network parameters are, of course, interrelated. Reed and Wheeler [4] gives the scattering parameters in terms of the odd and even reflection coefficients, Γ_e and Γ_o, and the odd and even transmission coefficients, T_e and T_o, as

$$S_{11} = \tfrac{1}{2}(\Gamma_e + \Gamma_o) \tag{1.45}$$

$$S_{12} = \tfrac{1}{2}(\Gamma_e - \Gamma_o) \tag{1.46}$$

$$S_{13} = \tfrac{1}{2}(T_e + T_o) \tag{1.47}$$

$$S_{14} = \tfrac{1}{2}(T_e - T_o) \tag{1.48}$$

The reflection and transmission coefficients are in turn connected to the cascade matrix parameters through

$$\Gamma = \frac{A + B - C - D}{A + B + C + D} \tag{1.49}$$

$$T = \frac{2}{A + B + C + D} \tag{1.50}$$

1.3 DEFINITION OF COUPLER PARAMETERS

The directional coupler parameters are defined in terms of the general network shown in Figure 1.3. The scattering matrix properties of the four-port can be described as

S_{11} = reflection coefficient at port i, i = 1–4;
S_{21} = coupling coefficient;
S_{31} = isolation;
S_{41} = transmission coefficient.

Being a symmetrical network, the ports can be rotated with retention of their power flow function. That is, if port 4 is made the input port, port 1 will be the through port, port 2 isolated, and port 3 coupled. Thus, the scattering matrix of the coupler can be written as

$$[S] = \begin{bmatrix} 0 & c & 0 & t \\ c & 0 & t & 0 \\ 0 & t & 0 & c \\ t & 0 & c & 0 \end{bmatrix} \tag{1.51}$$

A minor clarification of the parameter definitions is necessary. If a power divider is being constructed, there is no question of rotating the function of the ports.

Referring to Figure 1.3, the isolated port, 3, is then usually not even available physically, and for instance, the power could be split half way between ports 2 and 4. In the event of a mismatch at port 4, however, port 2 now becomes the isolated port to driven port 4 (the reflected power), and port 1 is the through port. It is clear that there must be a fourth port to absorb the power that would be coupled into the coupled port. (That this must be the case will be shown analytically in Chapter 2).

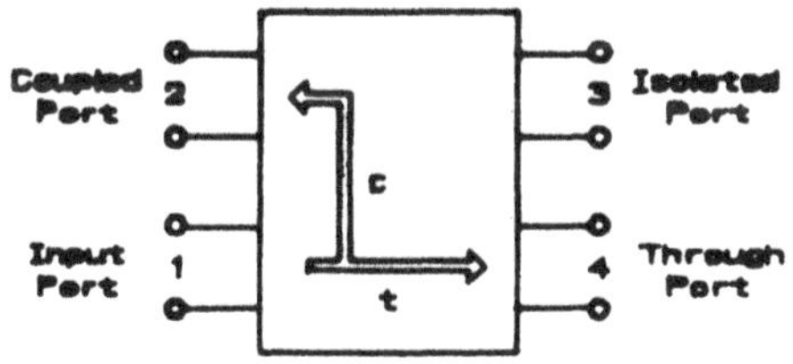

Figure 1.3 Definition of directional coupler parameters.

In the frequency response of the -3 dB coupler shown in Figure 1.4(a), the two coupling coefficients are nominally the same at the center frequency of the coupler. Away from the center frequency the coupling drops off as the frequency is increased or decreased, while the amount of power to the through port increases.

In order to extend the useful bandwidth of the coupler, which for argument's sake might be defined as the band of frequencies over which the coupling is within -3 ($\pm\frac{1}{2}$) dB, the coupling is made -2.5 dB, as shown in Figure 1.4(b). The coupled port now handles more power than the through port at band center. This is not a contradiction; in all cases port 2 will still be referred to as the coupled port because of the mechanism rather than the value associated with the specific scattering parameter. Note the increase in bandwidth in the second case.

The isolated port is completely isolated at all frequencies for coupled-line couplers, but this is only the case at bandcenter for branch-line couplers. In practice, the signals can never cancel completely in the isolated port and, in fact, can be of the same order of magnitude as the coupled signal, especially in the case of couplers with a low coupling coefficient. In such cases it is useful to use the ratio of power in the isolated port to that in the coupled port as a measure of the extent to which the coupler will function as a coupler. This ratio is termed the directivity of the coupler and is measured in dB.

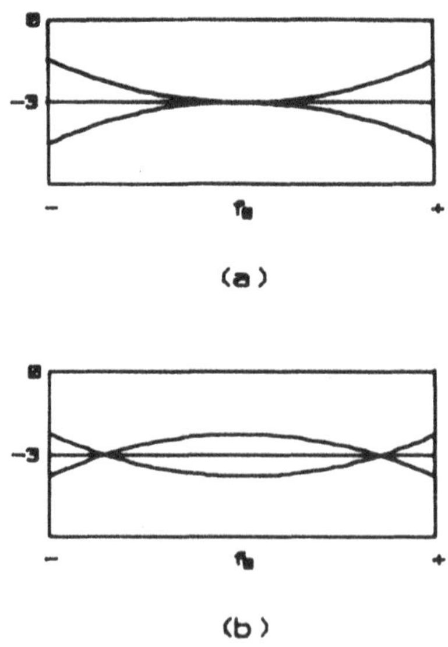

Figure 1.4 Response of coupling and transmission *versus* frequency. In (a) the coupling coefficient is -3 dB and -2.5 dB in (b).

The coupling characteristic shown in Figure 1.4 is band-limited, because the coupling mechanism is a resonance phenomenon with a certain periodicity. The form of the frequency *versus* coupling or transmission response is roughly sinusoidal, and the application obviously is limited by the bandwidth. We will show in later chapters that by cascading a number of sections with the proper coupling amplitudes, the bandwidth of the coupler can be greatly extended, to cover a decade in frequency or more.

1.4 PHYSICAL STRUCTURES

In branch-line couplers the unique relationship between phase and amplitude of the waves in the device make the coupling action possible. Figure 1.5(a) shows schematically the layout of a −3 dB, 90° coupler, while a 180° rat-race coupler is shown in Figure 1.5(b). In Figure 1.5(c) the resistor in the Wilkinson power divider forms the fourth port.

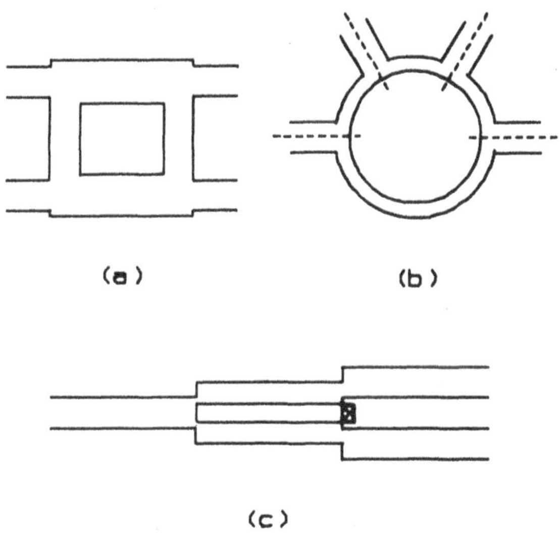

Figure 1.5 **Types of branch-line couplers: (a) 90° hybrid, (b) 180° rat-race coupler, and (c) a Wilkinson power divider.**

Coupled-line directional couplers consist of two or more parallel transmission lines coupled between a common pair of ground planes or its equivalent. Figure 1.6 shows the cross section through a variety of different types of line that can be used to realize coupled-line sections. The cross sections shown in Figure 1.6(a), (b), and (c) are stripline cross

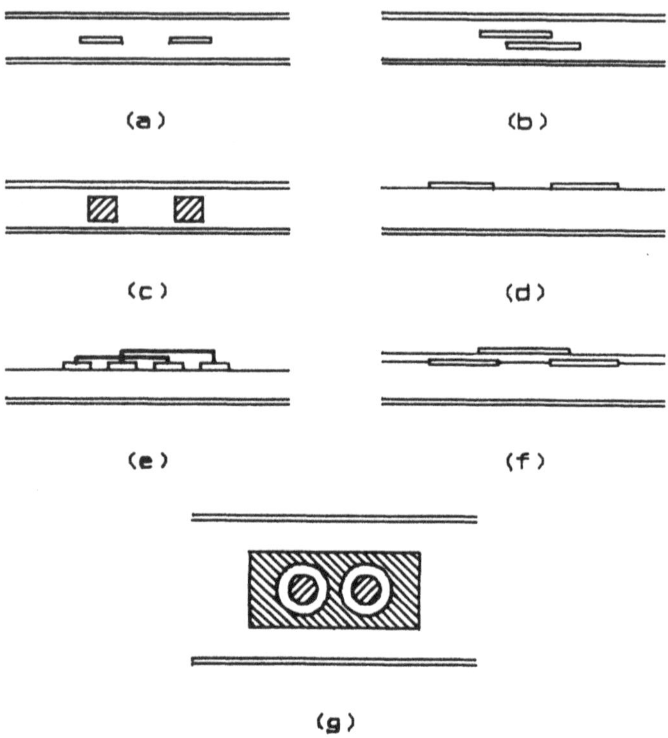

Figure 1.6 **Various cross sections commonly used in the realization of coupled-line directional couplers. (a), (b), and (c) are stripline configurations, while (d), (e), and (f) are microstrip constructions. The cross section shown in (g) is a reentrant section.**

sections, (a) being commonly used for cases of loose coupling, (b) for tighter coupling, and the cross section in (c) where high power is to be handled.

The cross section shown in (d) is a microstrip network for loose coupling, while (e) (the so-called Lange coupler) and (f) will provide coupling up to -3 dB. The section shown in (g) will give very tight coupling and handle high power. This has been termed a reentrant section [5].

Figure 1.7 illustrates the port connections of a simple single-section (a) and three-section (b) coupled-line coupler. Appendix B includes a computer program that calculates the dimensions of a coupler cross section making use of these standard structures. It must not be considered

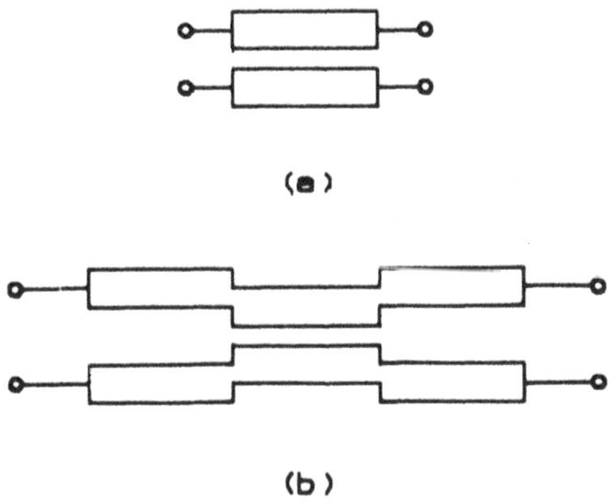

Figure 1.7 Single- and three-section coupled-line couplers.

the last word on calculation procedures for coupler cross sections but rather an aid to obtaining typical values that will serve as an indication of physical realizability.

1.5 ANALYSIS

As all design procedures at some stage or another involve an approximation to some desired property, it is worthwhile evaluating a design to check it for errors as well as its conceptual performance. The two computer programs in Appendix A run in BASIC and are intended for the analysis of three- and four-port symmetrical networks (SYMTHREE and SYMFOUR). The programs have been so constructed that the data input is in the same form as the theory in this text. In each section where an example is calculated, these programs have been used to calculate the network performance. The appendix contains full instructions on how to run the program, and the worked examples are analyzed.

REFERENCES

[1] J. L. Altman, *Microwave Circuits,* D. Van Nostrand, Princeton, NJ, 1964.

[2] J. A. G. Malherbe, *Microwave Transmission Line Filters,* Artech House, Dedham, MA, 1979.

[3] P. I. Richards, "Resistor–Transmission-Line Circuits," *Proc. IRE,* Vol. 36, February 1948, pp. 217–220.

[4] J. Reed and G. J. Wheeler, "A Method of Analysis of Symmetrical Four-Port Networks," *IRE Trans. Microwave Theory Tech.*, Vol. MTT-4, October 1956, pp. 246–252.

[5] S. B. Cohn, "The Re-entrant Cross Section and Wide-Band 3-dB Hybrid Couplers," *IEEE Trans. Microwave Theory Tech.*, Vol. MTT-11, July 1963, pp. 254–258.

Chapter 2
Branch-Line Directional Couplers and Power Dividers

2.1 INTRODUCTION

The use of connected-line directional couplers has already been mentioned in the introductory chapter. In this chapter the properties of connected-line couplers are analyzed by means of the two different analytical methods discussed in Chapter 1, namely by direct S-parameter synthesis and by means of odd and even network analysis.

The S-parameter synthesis procedure is not limited to directional couplers but can be applied to connected-line networks in general. Unfortunately, it becomes extremely complicated when applied to the synthesis of multisection networks, in which case analysis by odd and even networks will be used to find an equivalent circuit, after which the desired network is synthesized.

2.2 SCATTERING PARAMETER SYNTHESIS

The first step in scattering parameter synthesis is to obtain the scattering parameter matrix that describes the circuit to be synthesized. Thus, the method begins without knowledge of the physical nature of the resultant circuit or, indeed, knowledge of the realizability of the network.

For the moment it is assumed that the S-matrix is known; the procedure for obtaining a scattering matrix that will guarantee a specific performance will be discussed at a later stage. The first step is to convert the scattering matrix to an admittance matrix. This changes the known values specifying the performance of the circuit to a set of parameters specifying the network properties. Thus, from (1.2):

$$[y] = (I - S)(I + S)^{-1} \tag{2.1}$$

where $[I]$ is the unity matrix. It is now necessary to convert the admittance matrix into a network. The network to be synthesized will consist only of line sections (a multiple of an odd number of quarter-wavelength long sections, shunt connected only at the numbered ports).

Consider a network with N ports, and, let us say, the admittance parameter, y_{22}, needs to be determined. All the ports except port 2 are short circuited, and the admittance looking into port 2 obtained. If all the elements of the network are multiples of one-quarter wavelength, the admittance at port 2 must be zero, as all the input admittances of the lines connected to the port will be zero. This will be true for all the ports, so that it can be stated that the admittance matrix will have zero values along the diagonal, y_{ii}.

A further property of the admittance matrix becomes obvious when the mutual admittance between ports i and j is considered. Once again all the ports are short-circuited, except for the two ports under discussion, so that any line sections connected from i or j to any other port will have no effect on the mutual admittance y_{ij}. Obviously, this corresponds exactly to the condition that would have existed if the only element present were the one connecting ports i and j. We can therefore make the statement that the element values y_{ij}, y_{ji} represent a transmission line section connecting ports i and j, similar to a single, free-line section. Clearly, more information is needed on the properties of such sections of line.

Figure 2.1 shows two line sections, of length $\lambda/4$ and $3\lambda/4$ respectively, with characteristic impedance (admittance) of Z_1 (Y_1); as a matter of notation, the properties of circuits in a general impedance environment of Z_0, usually 50 Ω, will be considered. The impedances (admittances) are normalized as z_0 or y_0 to Z_0. The admittance matrix of a length of line l is given by

$$[Y] = \begin{bmatrix} -jY_1 \cot\beta l & jY_1 \csc\beta l \\ jY_1 \csc\beta l & -jY_1 \cot\beta l \end{bmatrix} \tag{2.2}$$

Consider the two unit elements of lengths $\lambda/4$ and $3\lambda/4$ shown in Figure 2.1. At band center, $\beta l = \pi/2$, and the two admittance matrices simplify to, respectively,

$$[Y] = j\begin{bmatrix} 0 & Y_1 \\ Y_1 & 0 \end{bmatrix} \tag{2.3}$$

and

$$[Y] = -j\begin{bmatrix} 0 & Y_1 \\ Y_1 & 0 \end{bmatrix} \tag{2.4}$$

The synthesis procedure will now be illustrated by means of a number of examples.

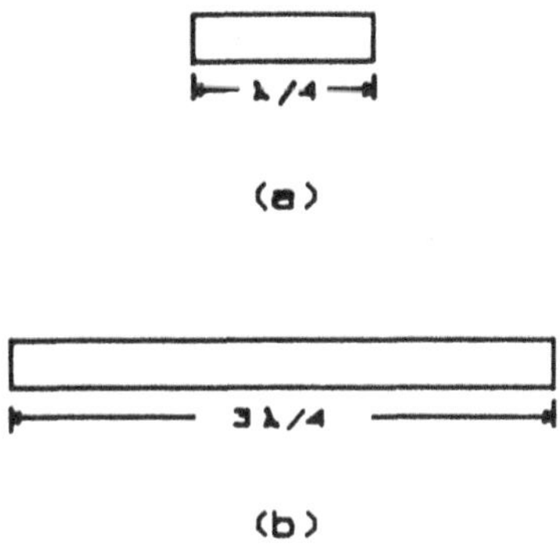

Figure 2.1 Unit elements (a) one-quarter and (b) three-quarters wavelengths long.

2.2.1 Synthesis of a 180° Hybrid Coupler

The hybrid coupler to be synthesized must split the input power equally between two ports that are out of phase. All ports must be matched, and if the port functions are defined in the usual way, the following scattering parameters can be written down by inspection [1]:

$$\begin{aligned} S_{11} &= S_{22} = S_{33} = S_{44} = 0 \text{ (match)} \\ S_{12} &= S_{21} = \sqrt{\tfrac{1}{2}} \text{ (half the power)} \\ S_{14} &= S_{41} = -\sqrt{\tfrac{1}{2}} \text{ (half the power and out of phase)} \\ S_{13} &= S_{31} = 0 \text{ (isolated)} \end{aligned} \tag{2.5}$$

Applying the unitary property to columns 1 and 2 shows that

$$S_{42} = 0$$

and from columns 2 and 4 follow

$$S_{32} \cdot S_{34}^{*} = \tfrac{1}{2}$$

Application of eq. (1.4) to row 3 gives

$$|S_{32}|^2 + |S_{34}|^2 = 1$$

and consequently we find that

$$S_{32} = S_{34} = \sqrt{\tfrac{1}{2}}$$

and by invoking symmetry,

$$S_{23} = S_{43} = \sqrt{\tfrac{1}{2}}$$

The scattering matrix is then given by

$$[S] = \sqrt{\tfrac{1}{2}}\begin{bmatrix} 0 & 1 & 0 & -1 \\ 1 & 0 & 1 & 0 \\ 0 & 1 & 0 & 1 \\ -1 & 0 & 1 & 0 \end{bmatrix} \tag{2.6}$$

The matrix in eq. (2.6) was obtained through the use of specified amplitudes only, except for the output phase power split. In effect this means that the ports have been specified to be physically coincident; in order to be realizable, the reference planes are all moved back by one-quarter wavelength, or $\pi/4$ radians.

The scattering matrix is now given by

$$[S] = j\sqrt{\tfrac{1}{2}}\begin{bmatrix} 0 & 1 & 0 & -1 \\ 1 & 0 & 1 & 0 \\ 0 & 1 & 0 & 1 \\ -1 & 0 & 1 & 0 \end{bmatrix} \tag{2.7}$$

The $[S]$ matrix is converted to a $[y]$ matrix,

$$[y] = j\sqrt{\tfrac{1}{2}}\begin{bmatrix} 0 & 1 & 0 & -1 \\ 1 & 0 & 1 & 0 \\ 0 & 1 & 0 & 1 \\ -1 & 0 & 1 & 0 \end{bmatrix} \tag{2.8}$$

Note that although it looks as if the two matrices are identical, this is pure coincidence and is not always the case. The matrix has been subdivided as

$$[y] = j\sqrt{\tfrac{1}{2}} \begin{bmatrix} 0 & 1 & 0 & -1 \\ 1 & 0 & 1 & 0 \\ 0 & 1 & 0 & 1 \\ 1 & 0 & 1 & 0 \end{bmatrix} \tag{2.9}$$

The matrix that would connect the ports 1 and 3 directly, shown in the box, has only zero elements, that is, $y_{13} = y_{31} = 0$. This means that there is no direct connection between ports 1 and 3. Similarly, the matrix elements that would connect ports 2 and 4 directly are zero, but the matrix connecting ports 1 and 2 is not zero (dotted lines):

$$[y] = j\sqrt{\tfrac{1}{2}} \begin{bmatrix} 0 & 1 & 0 & 0 \\ 1 & 0 & 0 & 0 \\ 0 & 0 & 0 & 0 \\ 0 & 0 & 0 & 0 \end{bmatrix} \tag{2.10}$$

This line is $\lambda/4$ long. Similarly, the matrix connecting ports 1 and 4 is

$$[y] = j\sqrt{\tfrac{1}{2}} \begin{bmatrix} 0 & 0 & 0 & -1 \\ 0 & 0 & 0 & 0 \\ 0 & 0 & 0 & 0 \\ -1 & 0 & 0 & 0 \end{bmatrix} \tag{2.11}$$

and $3\lambda/4$ long, and the connection between ports 2 and 3 is

$$[y] = j\sqrt{\tfrac{1}{2}} \begin{bmatrix} 0 & 0 & 0 & 0 \\ 0 & 0 & 1 & 0 \\ 0 & 1 & 0 & 0 \\ 0 & 0 & 0 & 0 \end{bmatrix} \tag{2.12}$$

This line is $\lambda/4$ long. The final connecting line is between ports 3 and 4, of length $\lambda/4$:

$$[y] = j\sqrt{\tfrac{1}{2}} \begin{bmatrix} 0 & 0 & 0 & 0 \\ 0 & 0 & 0 & 0 \\ 0 & 0 & 0 & 1 \\ 0 & 0 & 1 & 0 \end{bmatrix} \tag{2.13}$$

All other mutual admittance values are zero, as are all self-admittance values; this is necessary for connections of quarter-wavelength lines.

The network constructed from these submatrices is shown in Figure 2.2(a), or in the more familiar form of a rat-race coupler, in Figure 2.2(b). Note that all impedance levels inside the coupler are $y = \sqrt{\frac{1}{2}}$, $z = \sqrt{2}$, $Z = 70.7\ \Omega$. In the example, the power division factor was chosen to be $\frac{1}{2}$, but in fact any fraction could have been chosen without losing any generality of the procedure.

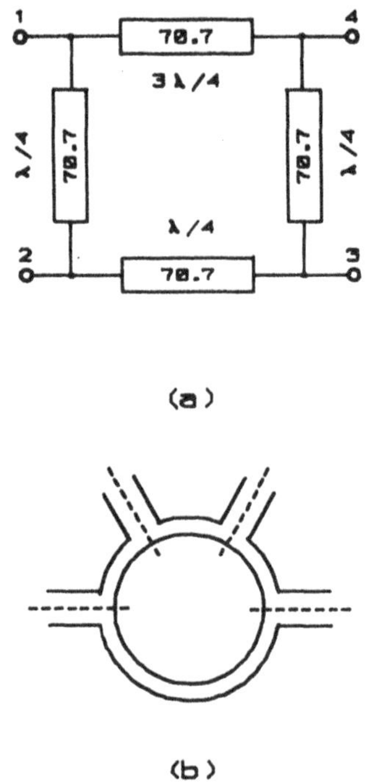

Figure 2.2 **(a) The network as synthesized. A common rat-race coupler is shown in (b).**

2.2.2 90° Branch-line Coupler

A similarly common (and popular) network is the 90° branch-line coupler with equal power split. Applying a procedure similar to the previous case gives the scattering matrix:

$$[S] = \sqrt{\tfrac{1}{2}}\begin{bmatrix} 0 & 0 & -j & 1 \\ 0 & 0 & 1 & -j \\ -j & 1 & 0 & 0 \\ 1 & -j & 0 & 0 \end{bmatrix} \tag{2.14}$$

If the reference planes are again shifted by $\lambda/4$:

$$[S] = -\sqrt{\tfrac{1}{2}}\begin{bmatrix} 0 & 0 & 1 & j \\ 0 & 0 & j & 1 \\ 1 & j & 0 & 0 \\ j & 1 & 0 & 0 \end{bmatrix} \tag{2.15}$$

and by applying eq. (2.1),

$$[y] = j\begin{bmatrix} 0 & 1 & 0 & \sqrt{2} \\ 1 & 0 & \sqrt{2} & 0 \\ 0 & \sqrt{2} & 0 & 1 \\ \sqrt{2} & 0 & 1 & 0 \end{bmatrix} \tag{2.16}$$

The network described by eq. (2.16) is shown in Figure 2.3; it has lines of $y = \sqrt{2}$ or $z = \sqrt{\tfrac{1}{2}}$, $Z = 35.4\ \Omega$ connecting ports 1 and 4 and 2 and 3, and lines of $Z = 50\ \Omega$ between 1 and 2 and 3 and 4. All lines are $\lambda/4$ long.

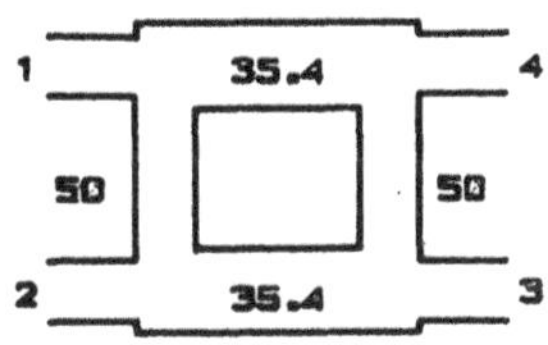

Figure 2.3 Physical layout of a 90° hybrid branch-line coupler.

2.2.3 Power Divider

This example is of a three-port network with special properties. The conditions are set that the power into port 1 must be equally divided between ports 2 and 3 and in phase. The network must, as usual, be matched. A further condition is that the two output ports must be isolated from each other. The scattering parameter matrix can be easily completed, and is found to be

$$[S] = -j\sqrt{\tfrac{1}{2}}\begin{bmatrix} 0 & 1 & 1 \\ 1 & 0 & 0 \\ 1 & 0 & 0 \end{bmatrix} \tag{2.17}$$

Note that a different procedure has been followed here. All the desired properties have been specified rather than obtained through deduction. The matrix in eq. (2.17) does not satisfy the unitary condition, implying that the network will not be lossless. Continuing as before yields a complex admittance matrix:

$$[y] = j\sqrt{\tfrac{1}{2}}\begin{bmatrix} 0 & 1 & 1 \\ 1 & 0 & 0 \\ 1 & 0 & 0 \end{bmatrix} + \tfrac{1}{2}\begin{bmatrix} 0 & 0 & 0 \\ 0 & -1 & 1 \\ 0 & 1 & -1 \end{bmatrix} \tag{2.18}$$

The real matrix describes a resistor of 100 Ω across ports 2 and 3. The resistor is needed to absorb the power that has to be dissipated if a mismatch occurs at one of the two output ports.

The Wilkinson power divider [2], is shown in Figure 2.4. This example indicates the power of the scattering matrix approach: it is essential for those cases where the physical form of the network is unknown.

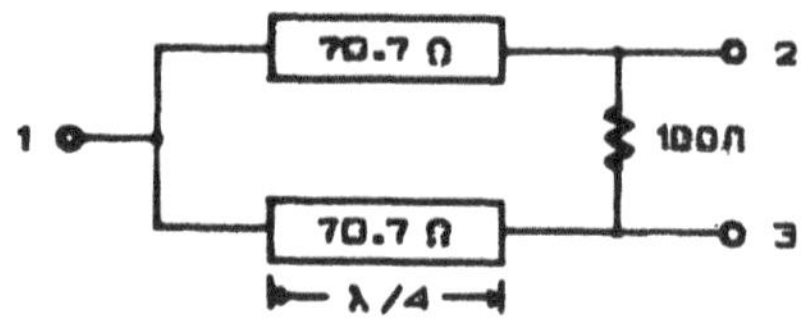

Figure 2.4 Physical layout of a Wilkinson power divider.

2.3 BRANCH-LINE HYBRID, ODD- AND EVEN-MODE ANALYSIS

For branch-line couplers of more than one section, the S-parameter synthesis procedure becomes too involved, and the odd- and even-mode analysis is used. Under these circumstances, the topology of the network must of course also be known. In this section, odd- and even-mode networks are used in analysis.

2.3.1 Even- and Odd-Mode Admittances

Figure 2.5(a) shows a branch-line network that is being fed in the even mode, and the odd-mode condition is shown in Figure 2.5(b). The resultant even- and odd-mode networks are shown in Figures 2.5(c) and 2.5(d), respectively. In the even-mode the branch lines become $\theta/2$ open-circuited stubs, while in the odd case, they become $\theta/2$ short-circuited stubs, of characteristic impedance z_1. The series lines in both cases remain unit elements of characteristic impedance z_2 and length θ. The input admittance of the even-mode network is given by

$$y_e = jy_1 \tan\tfrac{1}{2}\theta + y_2 \frac{(jy_1 \tan\frac{1}{2}\theta + 1) + jy_2 \tan\theta}{y_2 + j(jy_1 \tan\frac{1}{2}\theta + 1) \tan\theta} \tag{2.19}$$

whereas the odd-mode input admittance is given by

$$y_o = -jy_1 \cot\tfrac{1}{2}\theta + y_2 \frac{(-jy_1 \cot\frac{1}{2}\theta + 1) + jy_2 \tan\theta}{y_2 + j(-jy_1 \cot\frac{1}{2}\theta + 1) \tan\theta} \tag{2.20}$$

2.3.2 Performance at Band Center

At the center frequency $\theta = \pi/2$, $\tan\theta = \infty$, $\tan\theta/2 = \cot\theta/2 = 1$. The even- and odd-mode impedances become

$$y_e = jy_1 + \frac{y_2^2}{(jy_1 + 1)}$$

$$y_o = -jy_1 + \frac{y_2^2}{(-jy_1 + 1)}$$

Making use of (1.11),

$$S_{ij} = \frac{y_e - y_o}{(1 + y_e)(1 + y_o)}$$

the numerator of the equation is found as

$$2jy_1\left(1 - \frac{y_2^2}{1 + y_1^2}\right)$$

and the condition for isolation between ports 1 and 2 at center frequency is

$$y_2^2 - y_1^2 = 1 \tag{2.21}$$

The other scattering parameters can be obtained similarly by application of the same principles across the other plane of symmetry. The parameter S_{41} is obtained by noting that substituting y_2 for y_1 and *vice versa* yields the transmission coefficient. Substituting eq. (2.21) into eqs. (2.19), (2.20), and (1.11),

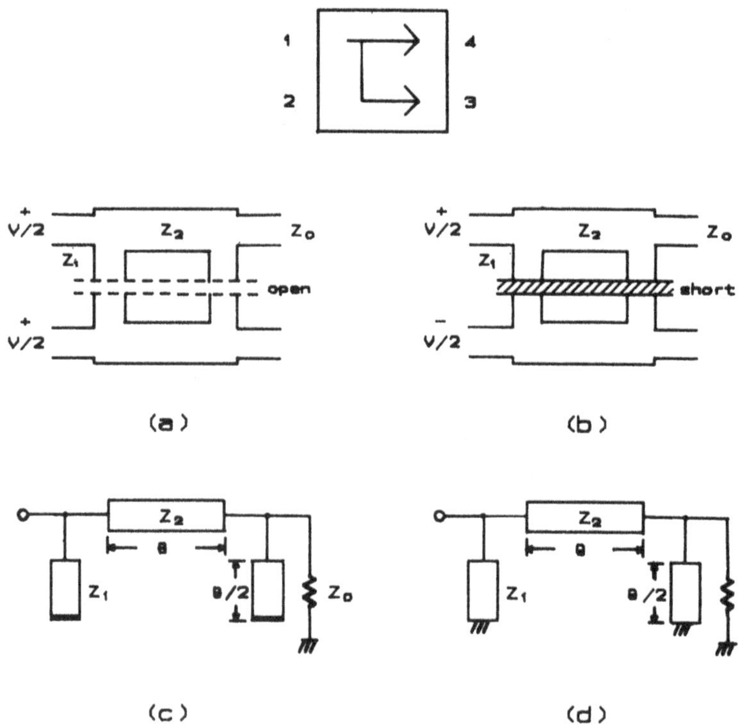

Figure 2.5 Even- (a) and odd-mode (b) networks for a symmetrical quarterwave network. The even and odd mode equivalent circuits are shown in (c) and (d).

$$S_{41} = j/y_2 = jz_2 \tag{2.22}$$

The reflection coefficient is obtained from (1.12):

$$S_{11} = \frac{1 - y_e y_o}{(1 + y_e)(1 + y_o)} \tag{2.23}$$

and if the condition (2.21) is substituted into (2.23),

$$S_{11} = 0$$

Finally, through the unitary property of the scattering parameters, at band center,

$$|S_{31}| = 1 - z_2^2 \tag{2.24}$$

Note that care has to be exercised in performing calculations and that the correct values of the odd and even admittances are used, as the networks are not diagonally symmetrical. The variation of coupled and through power with branch- and through-line impedances is shown in Figure 2.6.

As we will describe in Section 2.3.4, it becomes physically impractical to construct transmission lines of characteristic impedance beyond 110 or 120 Ω; taking physical realizability into consideration, this type of coupler can be constructed for coupling levels between about −3 dB and −6 dB.

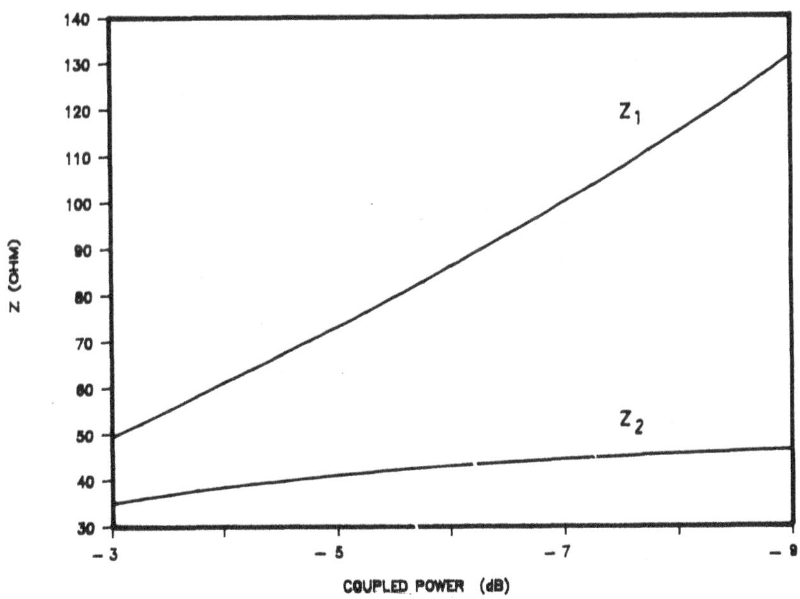

Figure 2.6 Variation of the branch-line and through-line impedances with coupled power level.

2.3.3 Variation of Parameters with Frequency

The variation of isolation, match, and transmission with frequency is obtained by substituting eqs. (2.19), (2.20), and (2.21) into (1.11) and (1.12), and the variation of coupled power with frequency is obtained by means of the principle of power conservation. These curves are shown in Figure 2.7. The results were obtained through use of the computer program SYMFOUR described in Appendix A; this specific case is treated as an example in Section A2.4.1.

Note that the branch-line coupler is matched only at band center. The typical -3 dB coupler shown is clearly limited to a bandwidth of not more than about 20 percent for coupling, if the criterion for coupling is taken to be a ± 0.5 dB variation about the nominal coupling level. If a criterion of 20 dB in isolation or return loss is considered, the bandwidth would only be 10 percent overall.

2.3.4 Limits on Physical Realization

In practice, there are limits to the impedance levels that can be realized in stripline and microstrip technology. On the one hand, a lower limit on impedance is set by the onset of moding, i.e., non-TEM-mode propagation. At the high impedance end, the limitation is due to the lines becoming so narrow as to drop off the substrate. For microstrip lines on alumina substrate ($\epsilon_r = 9.8$), the realizable impedances are bounded as $20\ \Omega \leq Z_0 \leq 120\ \Omega$; similar figures hold for berillia ($\epsilon_r = 6.6$) and fused quartz ($\epsilon_r = 3.8$). For typical striplines on plastic substrates ($\epsilon_r = 2.1$–2.6), realistic values for impedance are $40\ \Omega \leq Z_0 \leq 110\ \Omega$. Note that these are only approximate values. Designers may find that under specific circumstances much narrower lines and, thus, higher impedances can be realized by exercising the necessary care. As we will see in a following section, this limit on impedance is mainly a limiting factor in the realization of multisection couplers. In the case of single-section couplers the main application is for the -3 dB case, where the narrow linewidths pose no problem.

At the lower microwave frequencies, where the transmission lines are relatively narrow compared to their lengths, performance can be expected to be close to the theoretical values. As the frequency increases and the lines become shorter, the errors introduced by the junctions of the lines become more severe and have to be taken into consideration.

Celliers and Malherbe [3] present design curves for -3 dB branch-line couplers that compensate for the high frequency effects by changing the line impedances and lengths. Their method is based on a simultaneous

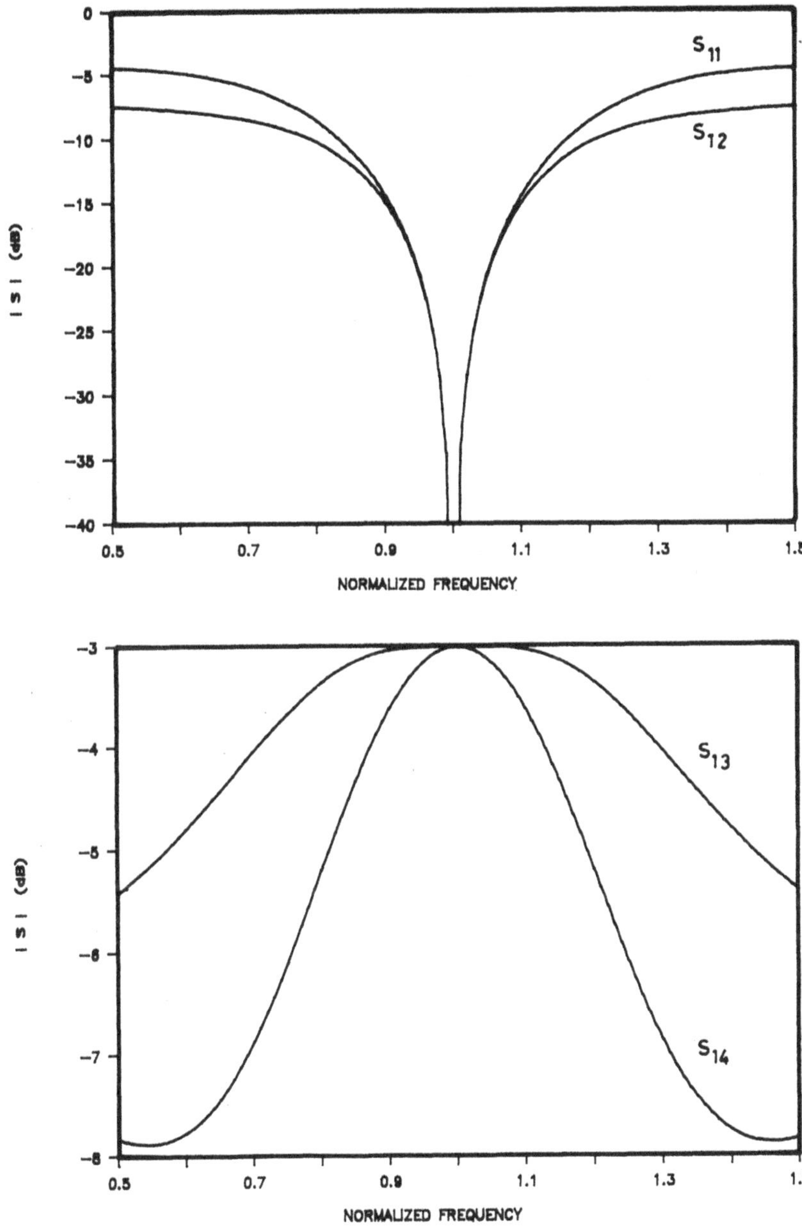

Figure 2.7 (a) Responses of the scattering parameters of a -3 dB branch-line coupler versus normalized frequency. (b) Shows an expanded view of the coupling and transmission.

optimization of the power split and isolation, and the design procedure is given here without comment. Referring to Figure 2.8, where the coupler dimensions are defined, calculate

$$D_i = \begin{cases} w_i + \dfrac{2b\ \ln 2}{\pi} + \dfrac{t}{\pi}\left[1 - \ln\left(\dfrac{2t}{b}\right)\right], & \text{for } \left(\dfrac{w_i}{b}\right) > 0.5 \\ b\,\dfrac{K(k)}{K(k')} + \dfrac{t}{\pi}\left[1 - \ln\left(\dfrac{2t}{b}\right)\right], & \text{otherwise} \end{cases} \tag{2.25}$$

where

$$i = 1, 2$$

$$k = \tanh\left(\frac{\pi w_i}{2b}\right) \tag{2.26}$$

$$k' = \sqrt{1 - k^2} \tag{2.27}$$

and $K(k)$ is an elliptic integral of the first kind.

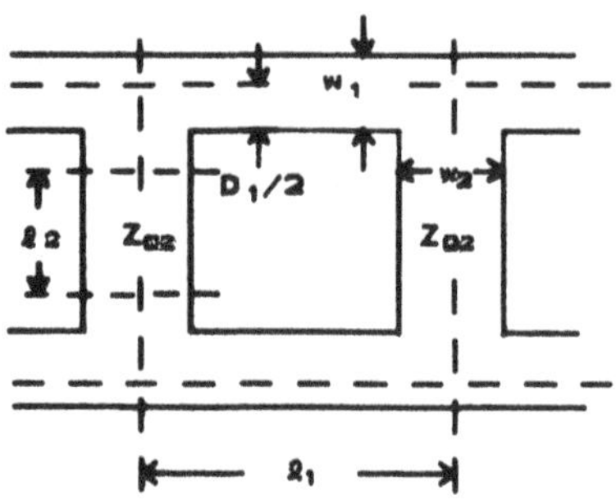

Figure 2.8 Geometry of branch-line coupler with optimized line lengths and impedance levels.

The normalized frequency is now calculated from

$$a = f \cdot b/0.538 \tag{2.28}$$

where f is the frequency in GHz, and b is the plate separation in mm. The impedances of lines Z_{01} are maintained at 50 Ω, but the characteristic impedance Z_{02} is read off the curve in Figure 2.9. New line lengths l_1 and

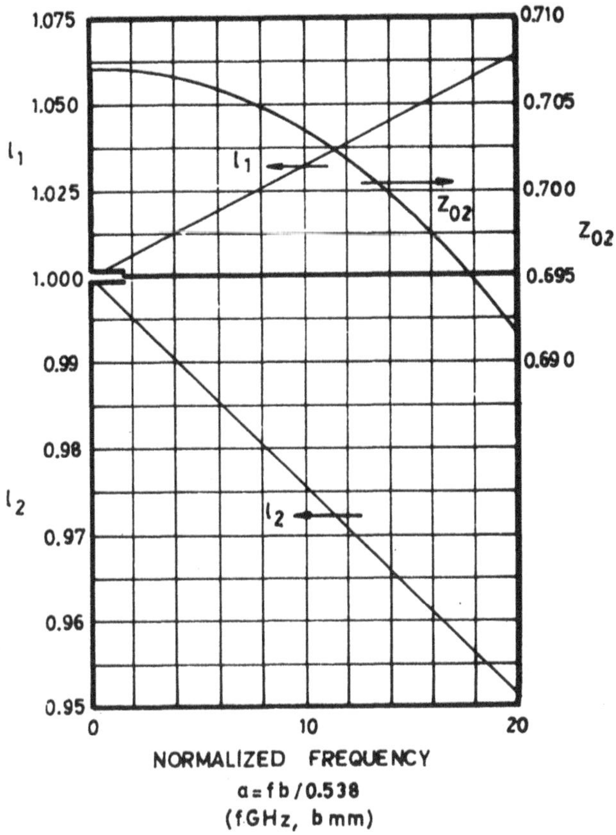

Figure 2.9 **Design curves for −3 dB branch-line couplers. (© 1985 IEEE, reprinted with permission from [3].)**

l_2 are also obtained; these lengths are given as a fraction of the resonant quarter-wavelength in Figure 2.9.

Note that the lines are not fed into the coupler in the conventional way, but that the input lines have been rotated through 90° in order to take advantage of the fact that the branch lines are of the same 50 Ω impedance as the feed lines.

2.4 RAT-RACE HYBRIDS

In the example of Section 2.2.1, the rat-race hybrid was synthesized from first principles. As was the case with the branch-line coupler, odd- and even-mode network analysis is more readily applied to the analysis of the network once it is in a known form. The rat-race hybrid has basically the same properties as the branch-line coupler, except that it has either a 0° or 180° phase shift between the two coupled and output ports.

2.4.1 Odd- and Even-Mode Analysis

The physical layout of an unequal split-ring rat-race coupler is shown in Figure 2.10(a), with the even- and odd-mode networks when driving ports 2 and 3 shown in 2.10(b) and 2.10(c), respectively. Applying the same arguments as earlier, and referring to Figures 2.10(d) and 2.10(e), the odd- and even-mode admittances are found as

$$y_e = jy_1 \tan\tfrac{1}{2}\theta + y_2 \frac{(jy_1 \tan 3\theta/2 + 1) + jy_2 \tan\theta}{y_2 + j(jy_1 \tan 3\theta/2 + 1) \tan \theta} \tag{2.29}$$

$$y_o = -jy_1 \cot\tfrac{1}{2}\theta + y_2 \frac{(-jy_1 \cot 3\theta/2 + 1) + jy_2 \tan\theta}{y_2 + j(-jy_1 \cot 3\theta/2 + 1) \tan\theta} \tag{2.30}$$

Unfortunately, the network possesses only one axis of symmetry, and although the even- and odd-mode admittances for ports 1–4 can be obtained, the previous equations cannot be used to calculate the scattering parameters S_{12}, S_{21}, S_{34}, and S_{43}. For this, the transmission line voltage equation (1.14) is used:

$$V_G = V_L(\cos\beta l + jZ_o/Z_L \sin\beta l)$$

Introducing the even-mode conditions:

$$\begin{aligned} V_{2e} &= V_{1e}(\cos\theta + y_{Le}/y_2 \sin\theta) \\ y_{Le} &= 1 + jy_1 \tan 3\theta/2 \end{aligned} \tag{2.31}$$

Substituting:

$$V_{1e} = \frac{V_{2e}}{\cos\theta + j(1 + jy_1 \tan 3\theta/2)/y_2 \sin\theta} \tag{2.32}$$

and similarly for the odd-mode network:

$$V_{1o} = \frac{V_{2o}}{\cos\theta + j(1 - jy_1 \cot 3\theta/2)/y_2 \sin\theta} \tag{2.33}$$

The even- and odd-mode generator voltages are given by

$$V_{2e} = \frac{\frac{1}{2}V}{1 + y_{2e}}, \qquad V_{2o} = \frac{\frac{1}{2}V}{1 + y_{2o}} \tag{2.34}$$

and the resultant generator voltage at port 2:

$$V_2 = V_{2e} + V_{2o} = \tfrac{1}{2}V\left(\frac{1}{1 + y_{2e}} + \frac{1}{1 + y_{2o}}\right) \tag{2.35}$$

The sum of the even and odd voltages, divided by the generator voltage V_2 gives

$$\begin{aligned} S_{12} &= (V_{1e} + V_{1o})/V_2 = S_{21} = S_{34} = S_{43} \\ &= \left[\frac{\left(\frac{1}{1 + y_{2e}}\right)}{\cos\theta + j(1 + jy_1 \tan 3\theta/2)/y_2 \sin\theta} \right. \\ &\quad \left. + \frac{\left(\frac{1}{1 + y_{2o}}\right)}{\cos\theta + j(1 - jy_1 \cot 3\theta/2)/y_2 \sin\theta}\right] \\ &\quad \div \left[\frac{1}{1 + y_{2e}} + \frac{1}{1 + y_{2o}}\right] \end{aligned} \tag{2.36}$$

2.4.2 Conditions at Band Center

At band center the condition for match at ports 2 and 3 is found as

$$y_1^2 + y_2^2 = 1 \tag{2.37}$$

and the transmission coefficients between the two ports as

$$S_{23} = S_{32} = jy_1 \tag{2.38}$$

If ports 1 and 4 are used in the odd- and even-mode analysis, the same condition as (2.37) is found for a match at the ports, but the scattering coefficients are found as

$$S_{14} = S_{41} = -jy_1 \tag{2.39}$$

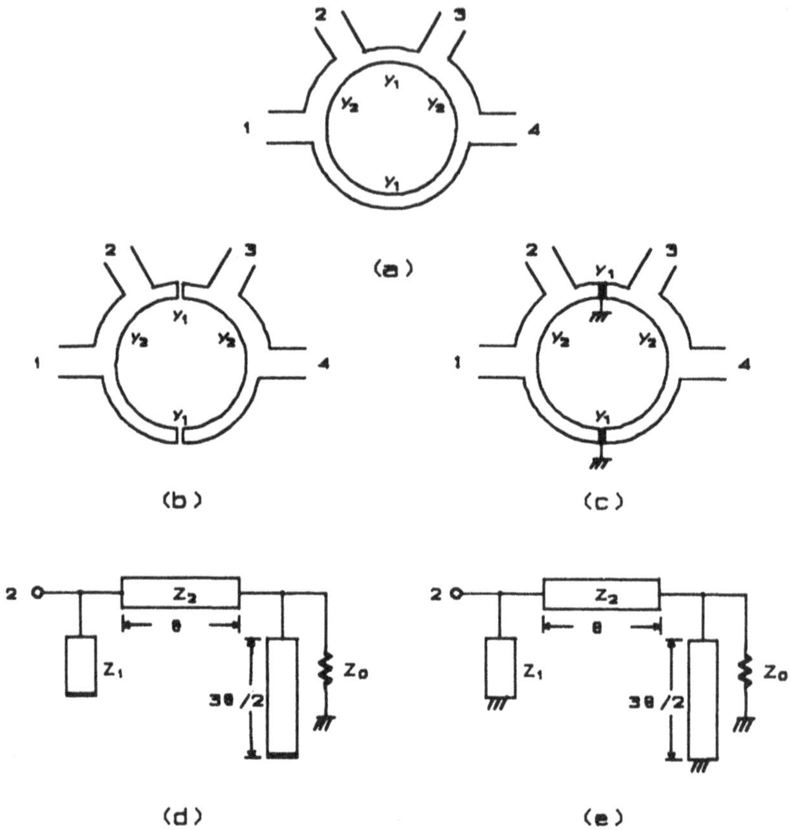

Figure 2.10 **(a) Physical layout of an unequal split-ring rat-race coupler, with the even- and odd-mode networks when driving ports 2 and 3 shown in (b) and (c), respectively. The odd- and even-mode admittance networks are shown in (d) and (e).**

which is in agreement with the physical properties of the network (i.e., a 180° phase shift between the two sets of ports).

From eq. (2.36) follows

$$S_{12} = jy_2 \tag{2.40}$$

The remaining scattering parameter is obtained as before from the scattering matrix:

$$[S] = \begin{bmatrix} S_{11} & S_{12} & S_{13} & S_{14} \\ S_{12} & S_{22} & S_{23} & S_{13} \\ S_{13} & S_{23} & S_{22} & S_{12} \\ S_{14} & S_{13} & S_{12} & S_{44} \end{bmatrix} \tag{2.41}$$

Application of the unitary property on the first row gives

$$S_{13} = 0 \tag{2.42}$$

Figure 2.11 shows the variation of line impedances with coupling level. For equal power split levels, the ring impedance is constant at 70.7 Ω. Due to one impedance rising to too high a level, hybrid ring couplers with a coupled power level of less than 6 or 7 dB are not practical.

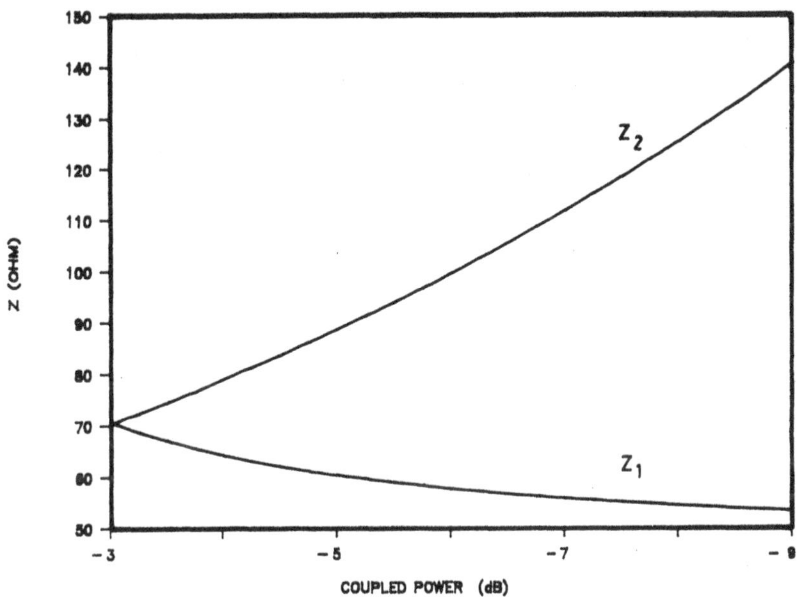

Figure 2.11 Variation of the line impedances of an unequal impedance rat-race coupler with coupled power level.

The variation of coupling parameters *versus* frequency for a coupling level of −3 dB is shown in Figure 2.12. It is once again clear that the coupler is not a wideband structure, a relative bandwidth of 20 percent

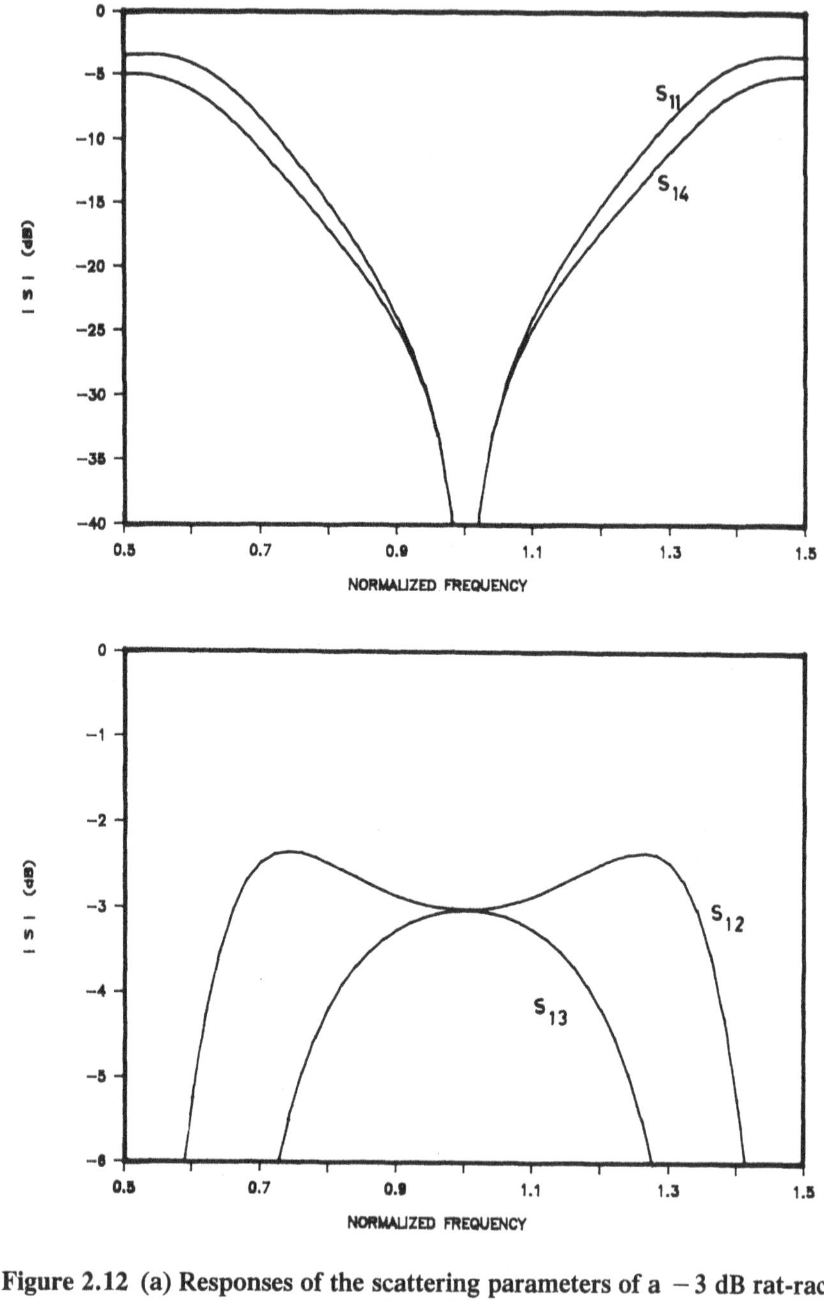

Figure 2.12 (a) Responses of the scattering parameters of a −3 dB rat-race coupler *versus* normalized frequency. (b) Shows an expanded view of the coupling and transmission.

being realistic. The coupler also suffers from the fact that with increased frequency the discontinuities of the junctions increasingly degrade the performance; and one way of solving this problem is by making the ring sections, respectively, $3\lambda/4$ and $5\lambda/4$ long. However, we can mount more components on the inside of the ring, as is frequently necessary (for instance, in the case of a balanced mixer). Increasing the line lengths unfortunately makes the network much more sensitive to frequency.

2.4.3 Stepped Hybrid-Ring Directional Coupler

Agrawal and Mikucki [4] have proposed a method to alleviate the serious limitations of realizability that prevail for coupling levels beyond −6 dB. The layout of the coupler is almost identical to that of the conventional hybrid-ring coupler, except that the $3\lambda/4$ line is divided into three sections, of characteristic admittance y_1, y_2, and y_1 respectively, as shown in Figure 2.13.

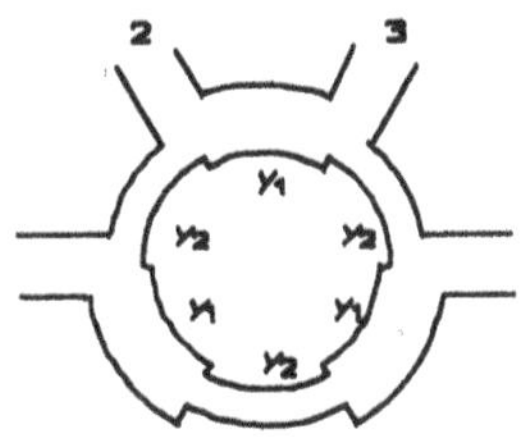

Figure 2.13 Stepped hybrid-ring directional coupler.

The analysis follows that of the conventional hybrid ring, except that the even- and odd-mode admittances at port 2 are now given by

$$y_{2e} = jy_1 \tan\tfrac{1}{2}\theta + y_2 \frac{y_{Le} + jy_2 \tan\theta}{y_2 + jy_{Le} \tan\theta} \tag{2.43}$$

where

$$y_{Le} = 1 + y_1 \frac{jy_2 \tan\tfrac{1}{2}\theta + jy_1 \tan\theta}{y_1 + j(jy_2 \tan\tfrac{1}{2}\theta) \tan\theta} \tag{2.44}$$

and

$$y_{2o} = -jy_1 \cot\tfrac{1}{2}\theta + y_2 \frac{y_{Lo} + jy_2 \tan\theta}{y_2 + jy_{Lo} \tan\theta} \tag{2.45}$$

where

$$y_{Lo} = 1 + y_1 \frac{-jy_2 \cot\frac{1}{2}\theta + jy_1 \tan\theta}{y_1 + j(-jy_2 \cot\frac{1}{2}\theta) \tan\theta} \tag{2.46}$$

The procedure to simplify the admittances at band center and to calculate the scattering parameters follows the same pattern as for the conventional hybrid ring and will not be repeated here. The coupler has one further distinct difference from the conventional coupler; the ports are not matched.

It is shown in [4] that the best simultaneous match at both ports 1 and 2 is obtained when

$$y_2^2 + \frac{y_1^3}{y_2} = 1 \tag{2.47}$$

Under these circumstances the scattering parameters at band center are given by

$$S_{11} = \frac{y_1^2(1 - y_1^2/y_2^2)}{4 + y_1^2(1 - y_1/y_2)^2} = S_{33} \tag{2.48}$$

$$S_{12} = \frac{-4jy_2}{4 + y_1^2(1 - y_1/y_2)^2} \tag{2.49}$$

$$S_{13} = \frac{-2y_1y_2(1 - y_1/y_2)}{4 + y_1^2(1 - y_1/y_2)^2} \tag{2.50}$$

$$S_{14} = \frac{2jy_1(1 + y_1/y_2)}{4 + y_1^2(1 - y_1/y_2)^2} \tag{2.51}$$

$$S_{23} = \frac{-2jy_1(1 + y_1/y_2)}{4 + y_1^2(1 - y_1/y_2)^2} \tag{2.52}$$

The remaining S-parameters can be obtained from eqs. (2.48)–(2.52) from symmetry.

The power split ratio is given by the ratio between S_{12} and S_{14}:

$$\frac{S_{12}}{S_{14}} = \frac{2y_2}{y_1(1 + y_1/y_2)} \tag{2.53}$$

Properties

The stepped hybrid-ring coupler has the disadvantage of not having the ports matched, and consequently the isolation is degraded as well. Figure 2.14 shows the variation of match and isolation as a function of the power split ratio. Both parameters remain at about −20 dB for all power split ratios larger than 6 dB (−7 dB coupling).

The advantage of the stepped hybrid ring lies in the much lower impedance levels needed for it to achieve the necessary power split than for the conventional hybrid ring. This is illustrated in Table 2.1, where the impedance levels of the two types of couplers are compared.

The conventional hybrid ring can only be realized for power split ratios up to about 6 dB, while power split ratios up to 10 dB can be realized with the stepped ring. The bandwidth of the stepped ring hybrid is also about 20 percent [4].

Table 2.1 Characteristic Impedances for Stepped and Conventional Hybrid-Ring Couplers

Coupling Coefficient (dB)	*Power Split Ratio (dB)*	*Conventional*		*Stepped*	
		Z_1	Z_2	Z_1	Z_2
−3.0	0	70.7	70.7	70.7	70.7
−6.0	5.0	99.8	57.8	84.9	57.1
−8.6	8.0	135.2	53.8	102.3	53.5
−10.4	10.0	165.8	52.4	118.6	52.1
−15.1	15.0	285.6	50.8	181.6	50.5

2.4.4 Other Realizations

A number of variations on the hybrid-ring coupler must be mentioned. Rehnmark [5] presents a design procedure for microwave hybrids constructed in a balanced line, and specifically for the case where the $3\lambda/4$-section is replaced by a $\lambda/4$-section plus a phase reversal obtained by twisting the pair of lines; this is of course only possible in a twin-wire configuration, which places serious restrictions on its application.

March [6] describes an application where the $3\lambda/4$ line is replaced by a single coupled-line section, diagonally connected, with the free ter-

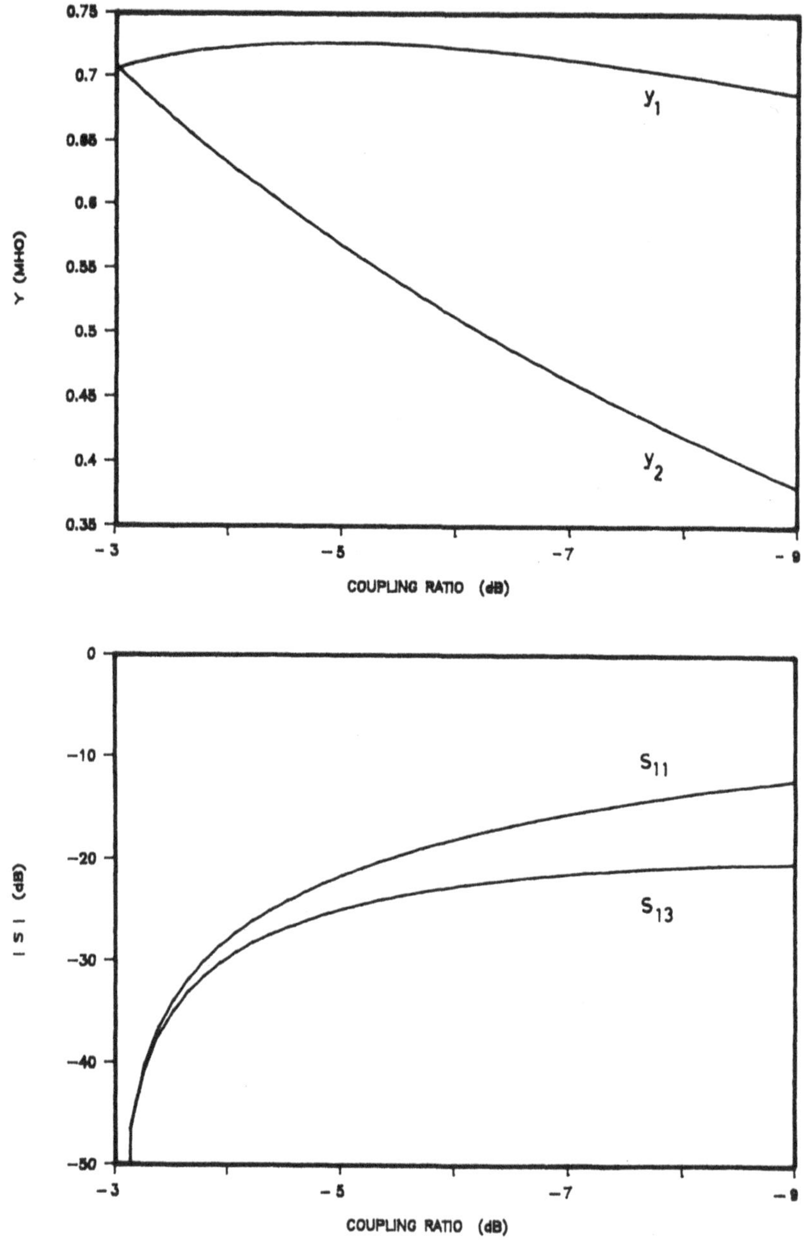

Figure 2.14 (a) Isolation and return loss *versus* coupling level for a stepped hybrid-ring coupler. The corresponding variation of admittance level is shown in (b).

minals shorted, as shown in Figure 2.15(a). The equivalent network model is shown in 2.15(c) and simplified in 2.15(d). The even-mode network for this section is shown in Figure 2.15(e). The even-mode input admittance of the section in Figure 2.15(e) is readily found as

$$y_{in} = -jy_{11}\cot\theta + jy_{12}\tan\tfrac{1}{2}\theta$$

which, at the band center, simplifies to

$$\begin{aligned} y_{in} &= jy_{12} = -\tfrac{1}{2}j(y_{11} - y_{11} - 2y_{12}) \\ &= -\tfrac{1}{2}j(y_{0e} - y_{0o}) \end{aligned} \tag{2.54}$$

Equating this to the even-mode input admittance of the $3\lambda/4$ section at midband:

$$y_1 = \tfrac{1}{2}(y_{0e} - y_{0o}) \tag{2.55}$$

In order for a coupled-line section to be matched to a given impedance, y_1, as will be seen in Section 4.2.4:

$$y_{0e}y_{0o} = y_1^2 \tag{2.56}$$

so that, after substitution:

$$y_{0e} = (\sqrt{2} - 1)y_1 \tag{2.57}$$

$$y_{0o} = (\sqrt{2} + 1)y_1 \tag{2.58}$$

The coupling and line widths for such a design are readily computed. The even- and odd-mode admittances of the coupled-line section vary more slowly with frequency than that of the conventional section, and consequently the coupling and other parameters of the coupler are less frequency dependent. Figure 2.16 shows how these parameters vary for a −3 dB coupler.

Kim and Naito [7] have shown that by considering the ring coupler as a general six-port device, the other two ports being connected at the quarter-wave points on the $3\lambda/4$ section, a ring coupler with four different impedances can be synthesized. This computer-optimized coupler has a useful bandwidth of 1.84 times that of the conventional design. The admittance values for a 50 percent bandwidth −3 dB coupler design are given in Figure 2.17. Its response is shown in Figure 2.18; refer to section A2.4.2 for a complete description of the analysis.

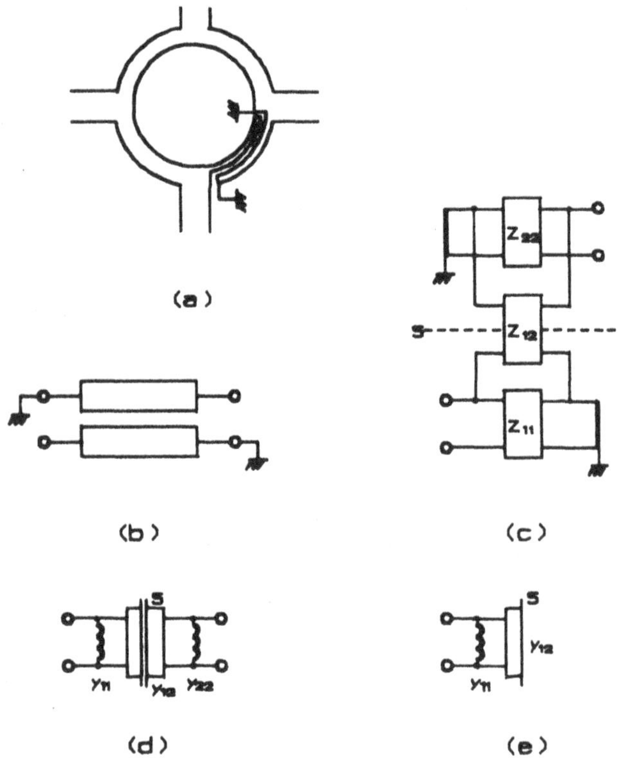

Figure 2.15 Hybrid-ring coupler using a pair of coupled lines (a). The coupled lines are shown in (b) and the network equivalent in (c). The simplified circuit is shown in (d) and the even mode network in (e).

2.5 POWER DIVIDERS

Power dividers can be distinguished from other couplers by their specific application. Although a variety of directional couplers will also serve as power dividers, the fact that an in-phase power split is desired usually makes the power divider a specific circuit. Under certain circumstances, a power combiner may be needed to combine the outputs from a number of devices, such as oscillators or amplifiers.

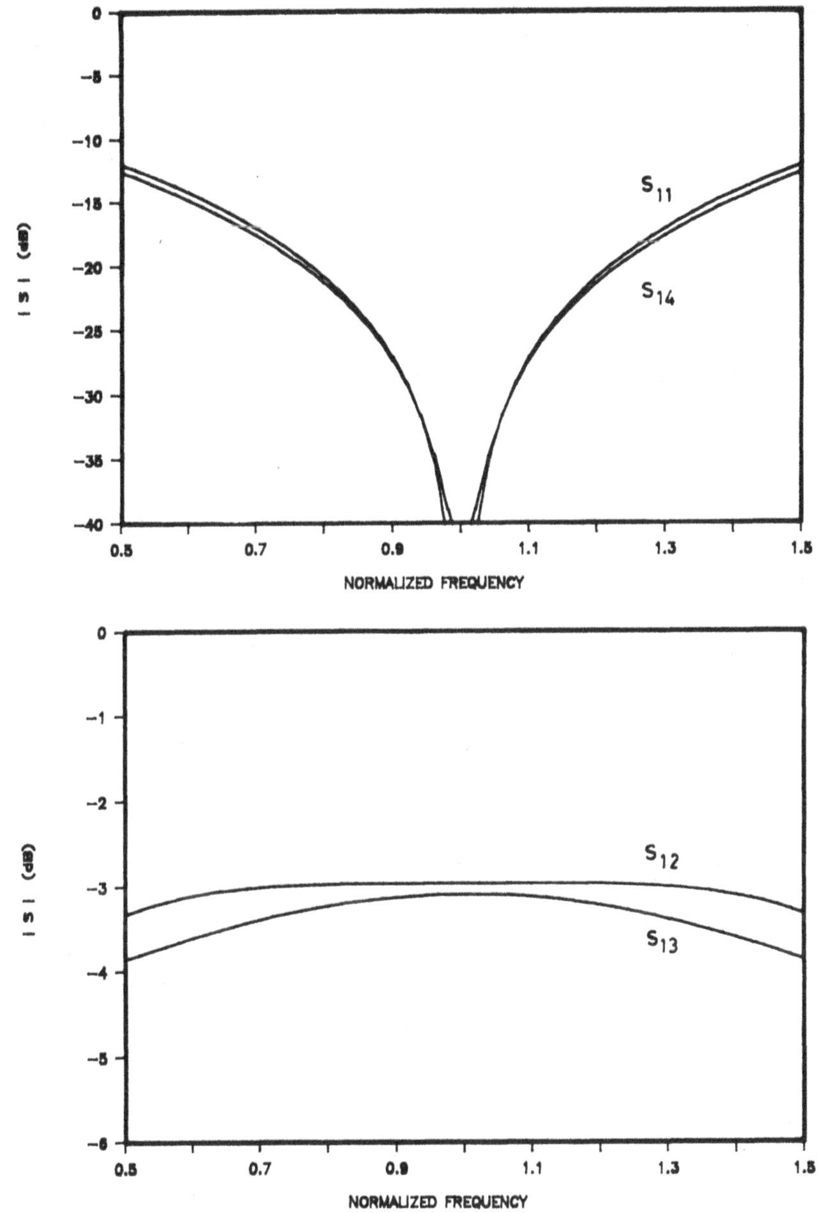

Figure 2.16 (a) Responses of the scattering parameters of a −3 dB hybrid ring with coupled-line section *versus* normalized frequency. (b) Shows an expanded view of the coupling and transmission.

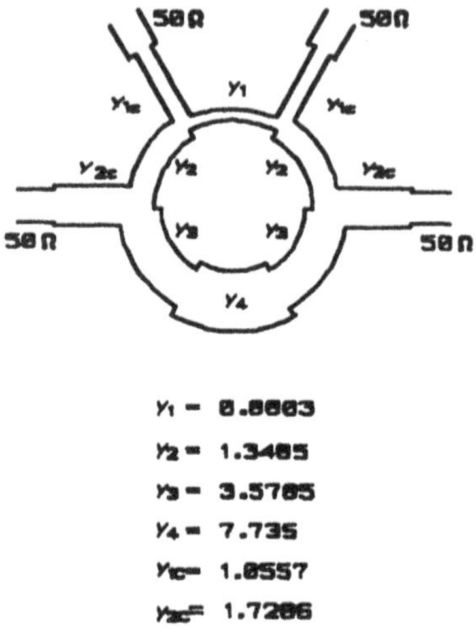

Figure 2.17 **Hybrid-ring coupler layout and admittance levels for 50 percent bandwidth design [7].**

2.5.1 Equal-Split Compensated Power Dividers

The sigle-section Wilkinson power divider was considered briefly in Section 2.2.3, with a power split between two ports. The original power divider described by Wilkinson [2] is in fact an N-way divider, but unfortunately needs a common connection point for a resistor from each of the outputs, making it impractical for planar networks with more than two output ports.

The bandwidth performance of the single-section coupler can be improved by compensation using an input transformer. Figure 2.19 shows such a power divider with its odd- and even-mode networks. The odd- and even-mode input admittances at band center at port 2 are given by

$$y_e = 2\, y_2^2/y_1^2 \tag{2.59}$$

$$y_o = 2/r \tag{2.60}$$

Making use of (1.11) gives the condition for isolation between ports 2 and 3 as

$$1 - y_e y_o = 0$$

$$1 - 2y_2^2/y_1^2 \cdot 2/r = 0 \tag{2.61}$$

$$(y_2/y_1)^2 = r/4$$

while the condition for match at 2 or 3, from (1.12) is

$$y_e - y_o = 0$$

$$(y_2/y_1)^2 = 1/r \tag{2.62}$$

Combining (2.61) and (2.62) gives the condition:

$$y_1 = \sqrt{2}y_2 \tag{2.63}$$

Note that this is also the condition for an input match. In order to improve the bandwidth performance of the power splitter, the properties of a two-section binomial transformer are considered. It has a maximally flat frequency response, and the admittance of the first section of the transformer is given by [8]:

$$Z_1 = Z_L^{1/4}\,Z_0^{3/4} \tag{2.64}$$

In the case of the uncompensated power divider, the line section of admittance y_2 can be considered to be a single-section transformer between load admittances of 1 and 2, and consequently, for the compensated power divider:

$$y_2 = (.5)^{1/4} = 0.8409$$

and from (2.63):

$$y_1 = 1.1892$$

The scattering parameters of compensated and uncompensated Wilkinson power dividers are compared in Figure 2.20. Figure 2.20(a) shows the coupled power level, together with the variation of S_{11}, which is the worst of the scattering parameters of the uncompensated device.

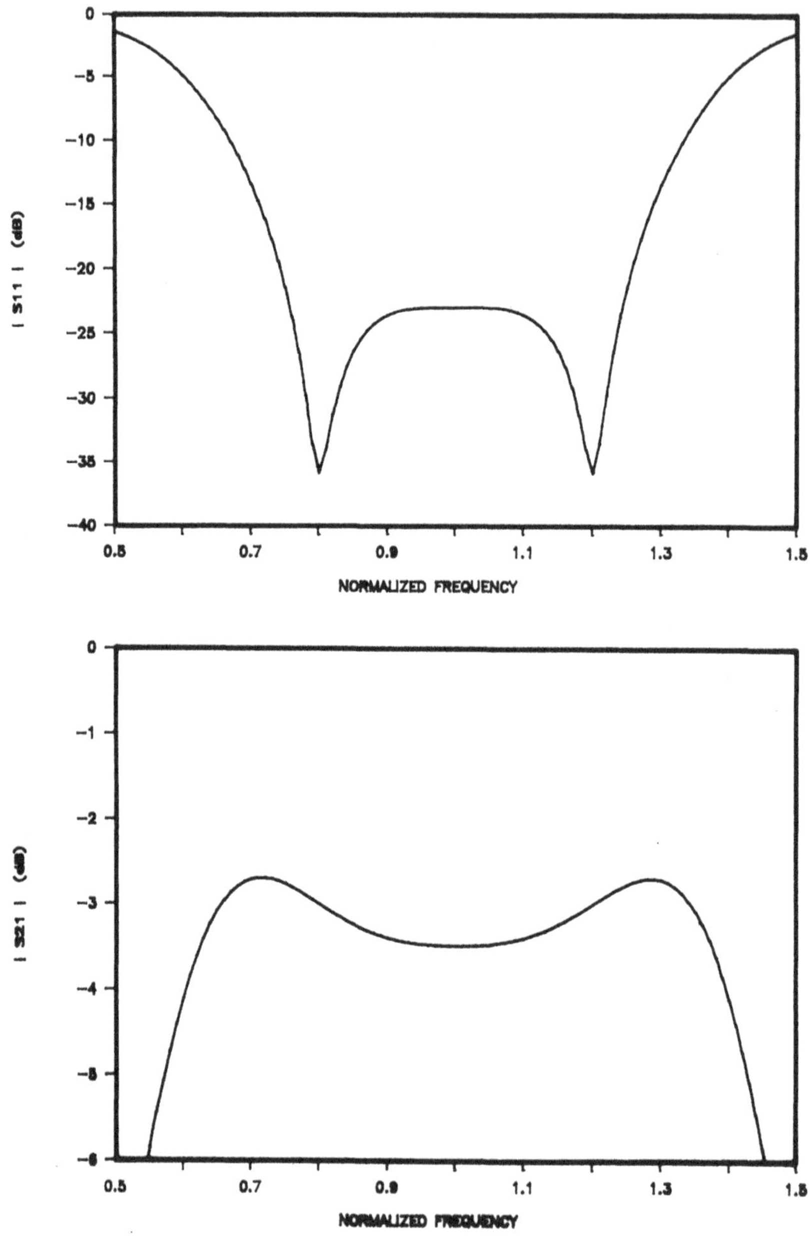

Figure 2.18 (a) Responses of the scattering parameters of an optimized −3 dB hybrid ring *versus* normalized frequency. (b) Shows an expanded view of the coupling.

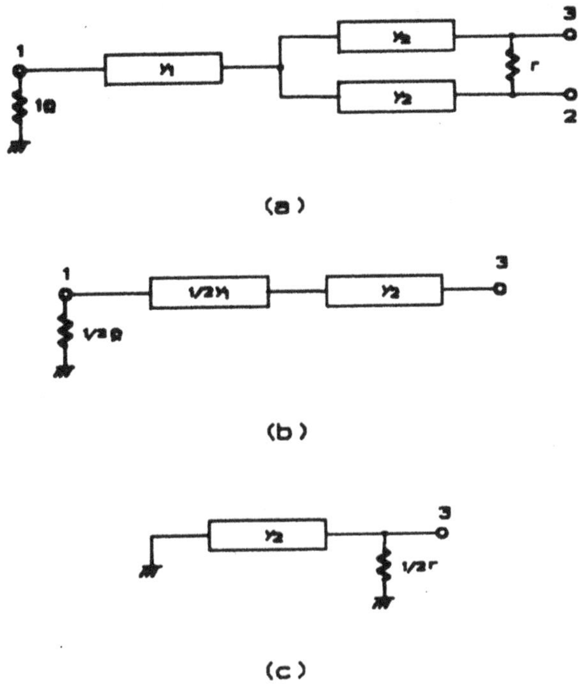

Figure 2.19 **(a) Schematic of a compensated Wilkinson power divider. The even- and odd-mode networks are shown in (b) and (c), respectively.**

The scattering parameters of the compensated power divider are shown in Figure 2.20(b).

If an acceptable level of return loss and isolation is taken to be -20 dB, the limiting parameters for the uncompensated coupler are the input return loss and isolation, both with a bandwidth of 36 percent. After compensation, the bandwidth is limited to 50 percent by the isolation.

2.5.2 Power Dividers with Unequal Power Split

The physical layout of a split-tee power divider [9] is shown in Figure 2.21(a). In order to split the output power between the two loads in the ratio of k^2, ports 2 and 3 have to be terminated, respectively, in admittances $1/k$ and k. These loads are transformed to the input such that port 1 is matched. Therefore,

$$y_{in} = \frac{y_1^2}{1/k} + \frac{y_2^2}{k} = 1 \qquad (2.65)$$

The input power must further divide in the correct ratio between the two branches, so that

$$\frac{y_2^2}{k} = k^2 \frac{y_1^2}{1/k} \qquad (2.66)$$

Substituting (2.65) into (2.64) yields

$$y_2^2 = k^2 y_1 \qquad (2.67)$$

$$y_1 = \frac{1}{\sqrt{k + k^3}} \qquad (2.68)$$

$$y_2 = \sqrt{\frac{k^3}{1 + k^2}} \qquad (2.69)$$

In the even-mode condition, the voltages along the two transmission lines are the same, and the terminal resistance at port 1 is divided according to the same ratio of admittances as in eq. (2.64). The two parts are found to be, respectively,

$$\frac{1}{1 + k^2} \quad \text{and} \quad \frac{k^2}{1 + k^2}$$

The odd- and even-mode circuits are shown in Figures 2.21(b) and 2.21(c). The even- and odd-mode admittances are given by

$$y_{e2} = \frac{y_1^2}{\left(\dfrac{1}{1 + k^2}\right)} = \frac{1}{k} \qquad (2.70)$$

$$y_{o2} = g_1 \qquad (2.71)$$

and for ports 2 and 3 to be isolated:

$$y_e - y_o = 0$$
$$g_1 = 1/k \qquad (2.72)$$

similarly, considering port 3, we have

$$g_2 = k \tag{2.73}$$

In order to match ports 2 and 3 to the system, each port is terminated in a quarter-wave transformer. Their characteristic admittances are given by

$$y_{t1} = \sqrt{k} \tag{2.74}$$

$$y_{t2} = 1/\sqrt{k} \tag{2.75}$$

The final circuit is shown in Figure 2.22.

The same principle as for the uncompensated power divider can be used to compensate the split-tee power divider. Referring to Figure 2.23, the admittances are given by,

$$y_0 = (k + 1/k)^{1/4} \tag{2.76}$$

$$y_1 = \frac{1}{[k^3 (1 + k^2)]^{1/4}} \tag{2.77}$$

$$y_2 = \left[\frac{k^5}{1 + k^2}\right]^{1/4} \tag{2.78}$$

$$y_{t1} = 1/\sqrt{k} \tag{2.79}$$

$$y_{t2} = \sqrt{k} \tag{2.80}$$

$$g = \frac{k}{1 + k^2} \tag{2.81}$$

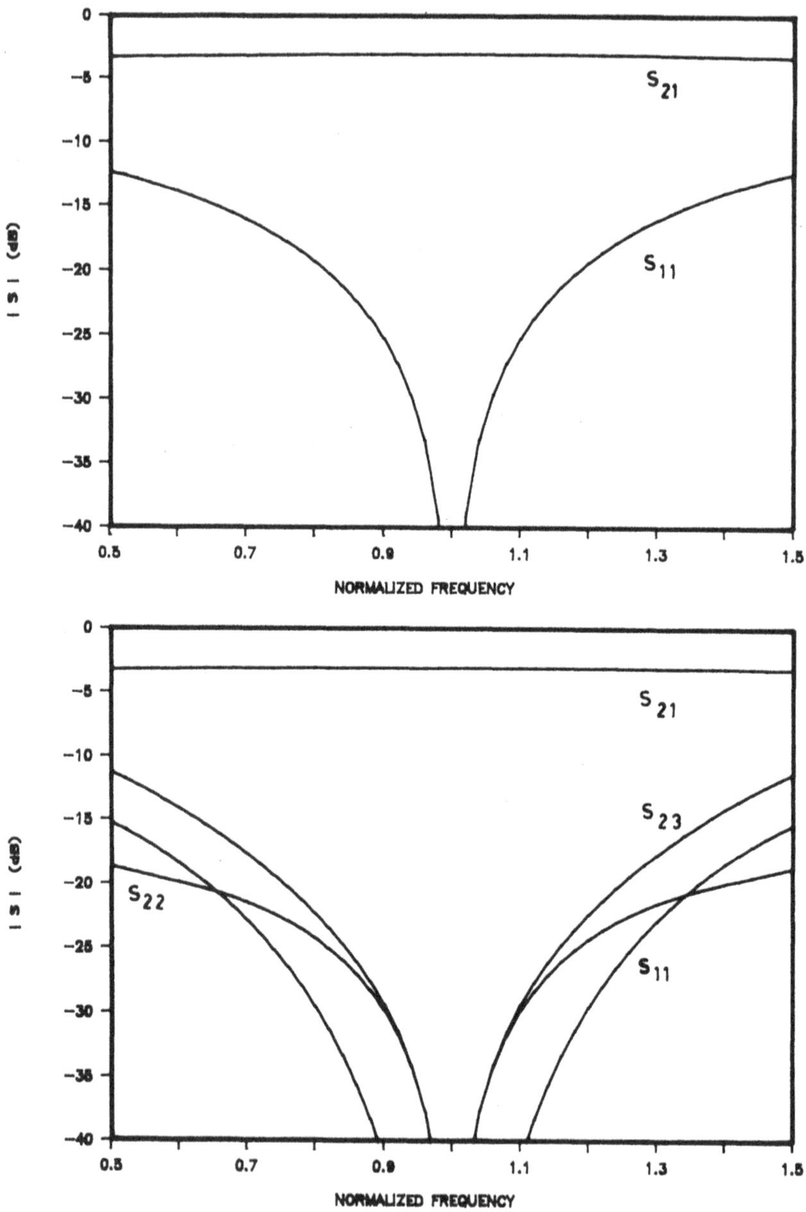

Figure 2.20 Comparison of the performance of compensated and uncompensatsed power dividers: (a) shows the coupled power and input return loss for the uncompensated device, and (b) the scattering parameters for the compensated power divider.

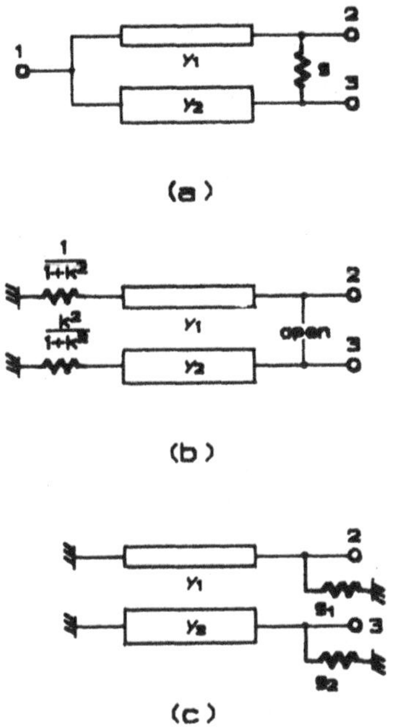

Figure 2.21 Physical layout: (a) of unequal-split power divider; the odd- and even-mode networks are shown in (b) and (c), respectively.

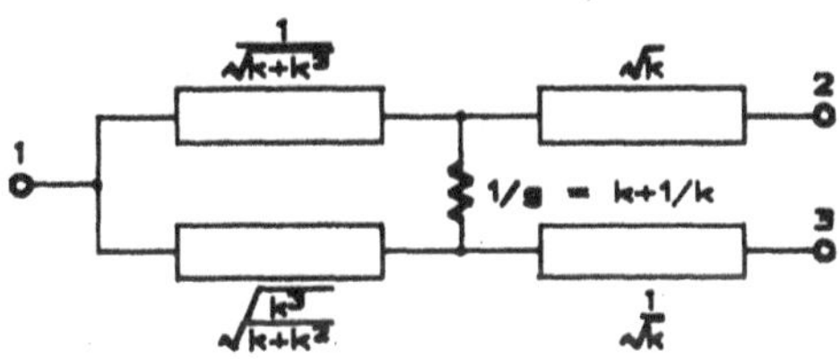

Figure 2.22 Split-tee power divider.

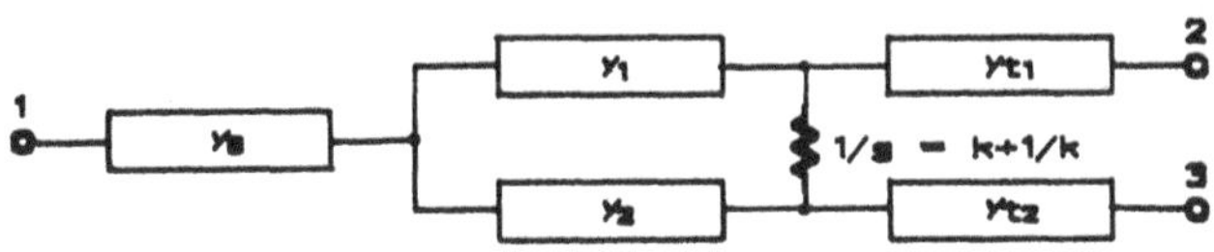

Figure 2.23 Compensated split-tee power divider.

REFERENCES

[1] J. L. Altman, *Microwave Circuits,* D. Van Nostrand, Princeton, NJ, 1964.

[2] E. J. Wilkinson, "An N-Way Hybrid Power Divider," *IRE Trans. Microwave Theory Tech.,* Vol. MTT-8, January 1965, pp. 116–118.

[3] A. F. Celliers and J. A. G. Malherbe, "Design Curves for −3 dB Branchline Couplers," *IEEE Trans. Microwave Theory Tech.,* Vol. MTT-33, November 1985, pp. 1226–1228.

[4] A. K. Agrawal and G. F. Mikucki, "A Printed-Circuit Hybrid-Ring for Arbitrary Power Divisions," *IEEE Trans. Microwave Theory Tech.,* Vol. MTT-34, December 1986, pp. 1401–1407.

[5] S. Rehnmark, "Wide-Band Balanced Line Microwave Hybrids," *IEEE Trans. Microwave Theory Tech.,* Vol. MTT-25, October 1977, pp. 825–830.

[6] S. March, "A Wideband Stripline Hybrid Ring," *IEEE Trans. Microwave Theory Tech.,* Vol. 16, June 1968, p. 361.

[7] D. I. Kim and Y. Naito, "Broad-Band Design of Improved Hybrid-Ring 3-dB Directional Couplers," *IEEE Trans. Microwave Theory Tech.,* Vol. MTT-30, November 1982, pp. 2040–2046.

[8] L. I. Parad and R. L. Moynihan, "Split-Tee Power Divider," *IEEE Trans. Microwave Theory Tech.,* Vol. MTT-13, January 1965, pp. 91–95.

[9] R. E. Collin, *Foundations for Microwave Engineering,* McGraw-Hill, New York, 1966.

Chapter 3
Wideband Connected-Line Couplers

3.1 INTRODUCTION

In the previous chapter, we employed a variety of structures to realize the coupling function. To a large extent these structures are band-limited, because they make use of a limited number of line sections and all the possible orders of freedom in the structure have been exploited. In this chapter, we describe directional couplers (and power dividers) that have very many more elements than are necessary solely to perform the coupling function at midband. These elements are introduced to increase the network's order of freedom so that it is possible to satisfy the additional requirements of a specific frequency response in one or more of the coupler parameters.

The general principles of network synthesis at microwave frequencies have been known for many years, but the wideband synthesis of multiport networks presents certain formidable problems. It is an extremely complex task to attempt the simultaneous synthesis of coupling, return loss, and directivity or isolation. In most cases, therefore two of these variables, at most, are synthesized at the same time, but fortunately the deviations from the ideal are usually only of academic interest. A serious problem, though, is the difficulty to realize many of the circuits in practice.

In this chapter, we treat two types of networks mainly; viz., the multisection branch-line coupler and the multisection in-line power divider. Some networks will use coupled lines. A substantial amount of the material in this chapter will not be of interest to the general reader and can be omitted. This includes Sections 3.2.1–3.2.3, and the derivations in Section 3.4.

3.2 WIDEBAND BRANCH-LINE COUPLERS

The procedure of synthesizing a directional coupler response that will be treated in this paragraph is typical of the type of argument used in network synthesis, and it consists of a number of steps. First, we study the properties of the physical structure, making use of cascade networks; this will lead to a function containing a number of unknowns that are determined by the network.

Subsequently, we fit the describing function to a polynomial such as Butterworth or Chebyschev, which will guarantee the desired parameter response. In the paragraphs that follow, we will use the synthesis procedure first described by Levy and Lind [1]. The procedure is valid only for symmetrical networks.

3.2.1 Matrix Properties of Symmetrical Couplers

The physical layout of a multisection coupler is shown in Figure 3.1. The symmetry is the same as in single-section branch-line couplers, a symmetry about two axes. Consequently, it is obvious that the complexity of the network will be simplified by analyzing the odd- and even-mode networks, shown in this case in Figure 3.2(a) and 3.2(b). The even-mode network consists of shunt-connected, open-circuited stubs spaced by unit elements, whereas for the odd-mode network, the stubs are short circuited.

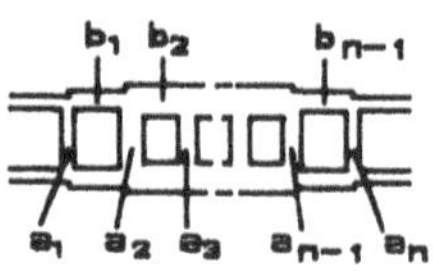

Figure 3.1 Multisection branch-line coupler.

The cascade matrix properties of networks were stated in Chapter 1. Specifically, the cascade matrix for each of the shunt capacitors with line admittance a_i in Figure 3.2(a) is given in terms of the Richards' variable by

$$[A] = \begin{bmatrix} 1 & 0 \\ a_i S & 1 \end{bmatrix} \tag{3.1}$$

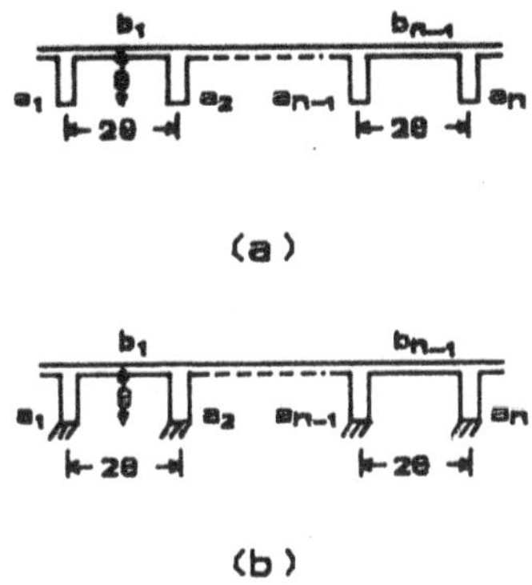

Figure 3.2 Odd- and even-mode networks for a multisection coupler.

where

$$S = \Sigma + j\Omega \tag{3.2}$$

$$\Omega = \tan\theta \tag{3.3}$$

The cascade matrix of each of the shunt inductors of admittance a_i in Figure 3.2(b) is

$$[A] = \begin{bmatrix} 1 & 0 \\ a_i/S & 1 \end{bmatrix} \tag{3.4}$$

The uniform lines connecting the branch lines are in effect cascades of two unit elements, each of characteristic admittance b_i, so that the cascade matrix of such a section is given by

$$[A] = \left\{ \frac{1}{\sqrt{1 - S^2}} \begin{bmatrix} 1 & S/b_i \\ b_i S & 1 \end{bmatrix} \right\}^2$$

$$= \frac{1}{1 - S^2} \begin{bmatrix} 1 + S^2 & 2S/b_i \\ 2b_i S & 1 + S^2 \end{bmatrix} \tag{3.5}$$

The cascade matrix of the even-mode network can now be obtained by simply multiplying the cascade matrices:

$$[A]_e = [a_1]_e \cdot [b_1]_e \cdot [a_2]_e \cdot [b_2]_e \cdot \ \ldots \ [a_{n-1}]_e \cdot [b_{n-1}]_e \cdot [a_n]_e$$

If this is done, the form of the even mode cascade matrix is found to be

$$[A]_e = \frac{1}{(1 - S^2)^{n-1}} \begin{bmatrix} A_{n-1}(S^2) & SB_{n-2}(S^2) \\ SC_{n-1}(S^2) & D_{n-1}(S^2) \end{bmatrix} \tag{3.6}$$

In (3.6) the subscripts indicate the order of the polynomials $A(S^2)$, $B(S^2)$, . . . , $D(S^2)$. Now it turns out that under the transform

$$S \rightarrow 1/S \tag{3.7}$$

we see that the cascade matrix for the shunt capacitors and shunt inductors are the transform of each other. Furthermore, the expression for the unit elements, eq. (3.5), is invariant to the transform, eq. (3.7), except that its sign changes. Consequently, if the cascade matrix of the even-mode network is given by

$$[A]_e = \begin{bmatrix} A_e(S) & B_e(S) \\ C_e(S) & D_e(S) \end{bmatrix} \tag{3.8}$$

then the cascade matrix of the odd mode is obtained by substituting $1/S$ for S in (3.8) and then multiplying the result with -1 for each unit element; namely, $(n - 1)$ times. The odd-mode cascade matrix is consequently given by

$$[A]_o = (-1)^{n-1} \begin{bmatrix} A_e(1/S) & B_e(1/S) \\ C_e(1/S) & D_e(1/S) \end{bmatrix} \tag{3.9}$$

Making use of the result (3.6) for the even-mode cascade matrix, we can therefore write the odd-mode cascade matrix as

$$[A]_o = \frac{(-1)^{n-1} S^{2(n-1)}}{(1 - S^2)^{n-1}} \begin{bmatrix} A_{n-1}(1/S^2) & S^{-1}B_{n-2}(1/S^2) \\ S^{-1}C_{n-1}(1/S^2) & D_{n-1}(1/S^2) \end{bmatrix} \tag{3.10}$$

From (1.36) the symmetry condition states that

$$A_{n-1}(1/S^2) = D_{n-1}(1/S^2) \tag{3.11}$$

3.2.2 The Synthesis Problem

The basis of the network synthesis procedure is to obtain an expression with a predetermined property for one of the network parameters: coupling, isolation, or reflection coefficient. The parameter applied to the synthesis procedure in this case is a combination of the ratios between the Γ and T at ports 1 and 2 or 3 and 4 for the even and odd modes. The expression for the ratio of reflection to transmission coefficients, from eqs. (1.49) and (1.50), is given by

$$\frac{\Gamma}{T} = \frac{1}{2}(B - C) \tag{3.12}$$

The expression [1] that will be used in the synthesis procedure is given by

$$F_{\pm} = \frac{1}{2}\left(\frac{\Gamma_e}{T_e} \pm \frac{\Gamma_o}{T_o}\right) \tag{3.13}$$

This function will be matched with the Butterworth or Chebyschev polynomials. It may be shown that the ratio Γ_e/T_e is of the form:

$$\frac{\Gamma_e}{T_e} = \frac{SP_{n-1}(S^2)}{(1 - S^2)^{n-1}} \tag{3.14}$$

where $P_{n-1}(S^2)$ is a polynomial in S^2 of order $(n - 1)$. Application of the transform (3.7) gives the odd-mode ratio:

$$\frac{\Gamma_o}{T_o} = \frac{(-1)^{n-1} S^{2n-3} P_{n-1}(1/S^2)}{(1 - S^2)^{n-1}} \tag{3.15}$$

We will now use these two ratios (3.14) and (3.15) to synthesize the response of the directional coupler; that is, the ratios will have the predetermined response properties and none of the primary coupler scattering parameters. The function (3.13) is written as

$$F_{\pm} = K\left[S \pm \frac{(-1)^{n-1}}{S}\right] \cdot \frac{P_{n-1}(S^2)}{(1 - S^2)^{n-1}} \tag{3.16}$$

3.2.2.1 Butterworth Response

A maximally flat characteristic of the ratio $F_{\pm}$ is obtained if all the zeros of the function are located at the center frequency of the shunt lines, $S = j$. Therefore,

$$P_{n-1}(S^2) = (1 + S^2)^{n-1} \tag{3.17}$$

and

$$F_{\pm} = K\left[S \pm \frac{(-1)^{n-1}}{S}\right] \cdot \frac{(1 + S^2)^{n-1}}{(1 - S^2)^{n-1}} \tag{3.18}$$

If (3.18) is split into its parts:

$$\frac{\Gamma_e}{T_e} = K\frac{S(S^2 + 1)^{n-1}}{(1 - S^2)^{n-1}} \tag{3.19}$$

$$\frac{\Gamma_o}{T_o} = K\frac{(-1)^{n-1} S^{2n-3} (1/S^2 + 1)^{n-1}}{(1 - S^2)^{n-1}} \tag{3.20}$$

Both Γ_e/T_e and Γ_o/T_o have $(n - 1)$ zeros at $S^2 = -1$, corresponding to those in $F_{\pm}$. Either of these functions can be used for synthesis, as we will show in Section 3.2.3.

3.2.2.2 Chebyschev Response

An equiripple response of the ratio $F_{\pm}$ results if the polynomial $P_{n-1}(S^2)$ is a Chebyschev polynomial. The expressions for the even- and odd-mode cascade matrices have to be rewritten to get them into the correct form. From eqs. (3.8) and (3.9), rewriting in terms of trigonometric functions, see eq. (1.42):

$$\begin{bmatrix} A_e & B_e \\ C_e & D_e \end{bmatrix} = \begin{bmatrix} A_{n-1}(-\cos 2\theta) & j\tan\theta B_{n-2}(\cos 2\theta) \\ j\tan\theta C_{n-1}(\cos 2\theta) & A_{n-1}(-\cos 2\theta) \end{bmatrix} \tag{3.21}$$

$$\begin{bmatrix} A_o & B_o \\ C_o & D_o \end{bmatrix} = \begin{bmatrix} A_{n-1}(\cos 2\theta) & j\cot\theta B_{n-2}(-\cos 2\theta) \\ j\cot\theta C_{n-1}(-\cos 2\theta) & A_{n-1}(\cos 2\theta) \end{bmatrix} \tag{3.22}$$

The variables of interest are $(B_e - C_e)$ and $(B_o - C_o)$, in order to form the ratios Γ/T. Inspection of eqs. (3.21) and (3.22) shows that the only differences between the even- and odd-cascade matrices are that the B_e and C_e are multiplied by $j\tan\theta$ while B_o and C_o are multiplied by $j\cot\theta$. Now, the even- and odd-cascade matrices are related through the transform (3.7), which means that the coefficients of the even and odd parts of the polynomials A, B, and C differ only in sign. For the symmetrical networks under consideration, B and C are either both even or both odd and, consequently, $(B_e - C_e)$ differs from $(B_o - C_o)$ only in the multiplying factor of $\pm\tan^2\theta$. The F_+ is therefore given by

$$F = \frac{1}{2}\left(\frac{\Gamma_e}{T_e} + \frac{\Gamma_o}{T_o}\right)$$

$$= \frac{1}{2}jK\tan\theta P_{n-1}\left(\frac{\cos2\theta}{\cos2\theta_c}\right) + \frac{1}{2}jK\cot\theta P_{n-1}\left(\frac{\cos2\theta}{\cos2\theta_c}\right)$$

$$= j\frac{K}{\sin2\theta}P_{n-1}\left(\frac{\cos2\theta}{\cos2\theta_c}\right) \tag{3.23}$$

where K is a constant, which, as earlier, controls the coupling, and $\cos\theta_c$ represents a scaled cutoff frequency. Consider the function:

$$P_L = 1 + |F_+|^2$$

$$= 1 + \frac{K^2P_{n-1}^2\left(\dfrac{x}{x_c}\right)}{(1 - x^2)} \tag{3.24}$$

where

$$x = \cos2\theta, \qquad x_c = \cos2\theta_c \tag{3.25}$$

In this case, K not only controls the coupling but also the ripple level of the Chebyschev function. Levy and Lind [1] point out that the function (3.24) is in the classic form of the transfer function for a cascade of $(n - 1)$ unit elements, and to obtain $(n - 2)$ ripples in the band $-x_c \leq x \leq x_c$:

$$P_{n-1}\left(\frac{x}{x_c}\right) = \frac{1}{2}(1 + \sqrt{1 - x_c^2})T_{n-1}\left(\frac{x}{x_c}\right)$$

$$- \frac{1}{2}(1 - \sqrt{1 - x_c^2})T_{n-3}\left(\frac{x}{x_c}\right) \quad (3.26)$$

where $T_n(x)$ is a Chebyschev polynomial of the first kind of degree n.

From (3.23) it can be seen that

$$\frac{\Gamma_e}{T_e} = jK \tan\theta P_{n-1}\left(\frac{x}{x_c}\right) \quad (3.27)$$

where

$$\frac{x}{x_c} = \left(\frac{1 + S^2}{1 - S^2}\right) \cdot \left(\frac{1 - S_c^2}{1 + S_c^2}\right) \quad (3.28)$$

It is obvious that, except for the origin and at infinity, Γ_o/T_o has the same zeros as Γ_e/T_e and that the responses specified by (3.27) and (3.28) will give a nearly equiripple response in the reflection coefficient and isolation. The unitary properties of both the even- and odd-mode networks ensure that

$$|\Gamma|^2 + |T|^2 = 1$$

so that

$$|T|^2 = \frac{1}{1 + |\Gamma_e/T_e|^2} \quad (3.29)$$

Substituting the values of Γ/T into (3.29) shows that at band center, both T_e and T_o are very close to unity over the passband. Consequently, Γ_e and Γ_o are very close to equiripple and also, by (1.45) and (1.46), the reflection coefficient and isolation of the coupler.

3.2.3 Synthesis Procedure

Once the functions Γ/T have been obtained, the synthesis of the desired even- or odd-mode network is straightforward though tedious. In essence, it boils down to obtaining an expression for Γ_e from

$$|\Gamma_e|^2 = \frac{|\Gamma_e/T_e|^2}{1 + |\Gamma_e/T_e|^2} \tag{3.30}$$

The numerator of (3.30) is always a perfect square, and the roots of the denominator are obtained numerically; only those poles lying in the left half of the S-plane are chosen. Once Γ_e has been obtained, the input impedance of the even-mode network is readily found from

$$Z_e = \frac{1 + \Gamma_e}{1 - \Gamma_e} \tag{3.31}$$

Levy and Lind [1] obtained extensive tables of prototypes for multisection branch-line couplers and fully describe the synthesis procedure mentioned earlier; further references and discussion of the synthesis procedure are available in (Malherbe [2], Ch. 6).

3.2.4 Prototypes

The branch-line coupler prototypes obtained by this procedure suffer from one serious disadvantage, extreme impedance levels. Although this does not present serious problems in waveguide construction, it does limit the applicability of practical circuits rather severely. Table 3.1 shows a selection of prototypes for Butterworth responses; Chebyschev prototypes are given in Table 3.2. Only prototypes with line characteristic admittances $y_0 \leq 0.333$, corresponding to impedances of less than 150 Ω, are realistic.

We see that this criterion allows only a number of prototypes with tighter than 4 dB coupling and a maximum of three branch lines. As a matter of interest, for the tightest (−2 dB) coupling with a Butterworth response and four branch lines, the admittance levels are, respectively, for the branch lines, 0.2102 and 1.4274; for the main lines the admittances are acceptable. For all prototypes with more numbers of lines, the admittance levels are even more severe. For the equiripple case, the realizability is slightly better but still unacceptable for $n > 3$.

In [3], Levy develops prototypes that make use of a Zolotarev polynomial rather than the Chebyschev or Butterworth approximations. Although there is some improvement in the impedance levels, the resultant branch-line impedance levels are still unrealizable.

Table 3.1 Element Values for Butterworth Branch-Line Couplers (© 1968 IEEE, reprinted with permission, from [1])

Coupling (dB)	*Band Edge Coupling (dB)*	*Relative Band-width*	*VSWR*	*Line Admittances*		
				Branch		*Main*
				a_1	a_2	b_1
4	3.7	1.33	1.12	0.3553	0.9974	1.2573
3	2.69	1.33	1.19	0.4140	1.3432	1.3774
2.5	2.24	1.30	1.22	0.4513	1.6188	1.4693
2	1.74	1.30	1.29	0.4941	2.0325	1.5996

Table 3.2 Element Values for Chebyschev Branch-Line Couplers (© 1968 IEEE, reprinted with permission, from [1])

Relative Band-width	*Coupling (dB)*	*Band Edge Coupling (dB)*	*Directivity (dB)*	*VSWR*	*Line Admittances*		
					Branch		*Main*
					a_1	a_2	b_1
	5	4.83	31.9	1.02	0.3174	0.7686	1.1874
1.22	4	3.83	31.2	1.03	0.3681	0.9729	1.2641
	3	2.83	30.4	1.05	0.4330	1.3103	1.3899
	5	4.32	19.3	1.10	0.3499	0.7087	1.2011
1.50	4	3.30	18.5	1.15	0.4130	0.8958	1.2880
	3	2.29	17.7	1.24	0.5007	1.2156	1.4391

The synthesis procedures described earlier do not specify the allowable immittance levels of the lines, and so they are optimal in a response sense only. Muraguchi et al. [4] set a limit on the allowable admittance level and then make use of a computer optimization routine to obtain a prototype that is optimal under an admittance constraint, thus sacrificing some performance in order to gain realizability. They consider realizable microstrip impedance levels to be 10–100 Ω for alumina substrates in the band 4–8 GHz, or 40–160 Ω, depending on the substrate thickness.

The acceptable limit of coupler performance was set at -3 ± 0.43 dB for the coupling imbalance and -20 dB for return loss and isolation. Table 3.3 shows the relevant parameters from [4] for a variety of couplers with three branch lines. The symbols in the tables have the following meanings:

BW_c	=	Bandwidth over which the coupling is better than -3 ± 0.43 dB
$BW_{R,I}$	=	Bandwidth over which both return loss and isolation are better than -20 dB
k_0	=	Coupling imbalance at the center frequency
S_{min}	=	Minimum return loss over $BW_{R,I}$
S_{max}	=	Maximum retrun loss over $BW_{R,I}$
$1/D$	=	Sampling interval over the frequency bandwidth in the optimization routine

Table 3.3 Element Values for Optimized Branch-Line Couplers (n = 3) (© 1983 IEEE, reprinted with permission, from [4])

	D	$BW_{R,I}$ (%)	BW_c (%)	k_0 (dB)	b_1 (Ω)	a_1 (Ω)	a_2 (Ω)
3-1	—	29	28	0	36.30	120.51	37.22
3-2	—	39	27	0.18	34.74	99.86	41.13
3-3	18	33	39	0.18	26.95	102.88	25.26
3-4	20	32	38	0.14	26.88	105.17	24.29
3-5	20	36	37	0.19	29.41	104.95	29.41
3-6	20	38	37	0.23	31.23	104.98	33.33
3-7	20	39	36	0.29	34.87	105.55	41.67
3-8	20	39	35	0.33	38.29	106.66	50.00
3-9	18	38	35	0.47	42.35	106.72	62.50
3-10	20	36	34	0.39	43.14	108.74	62.50
3-11	22	35	33	0.32	43.74	110.35	62.50
3-12	16	38	37	0.63	44.33	105.51	71.43
3-13	18	35	35	0.51	45.46	107.97	71.43
3-14	20	33	33	0.41	46.33	110.01	71.43
3-15	16	34	37	0.68	47.98	106.56	83.33
3-16	18	31	34	0.55	49.23	108.98	83.33
3-17	20	29	33	0.44	50.15	110.84	83.33

The large variety of possible solutions arise because a number of possible parameters can be held constant or at a predetermined level. The coupler 3-1 is the Butterworth prototype, while 3-2 has a Chebyschev response. Couplers 3-9, 3-10, and 3-11 all have a_2 preset at a level of 62.50 Ω, *etcetera*.

Table 3.4 shows similar prototypes for couplers with four branches. The impedance values of [4] in the a_1 and a_2 columns have been rounded.

Table 3.4 Element Values for Optimized Branch-Line Couplers ($n = 4$) (© 1983 IEEE, reprinted with permission, from [4])

	D	$BW_{R,I}$ (%)	BW_c (%)	k_0 (dB)	b_1 (Ω)	b_2 (Ω)	a_1 (Ω)	a_2 (Ω)
4-1	—	43	32	0	39.97	30.33	263.9	52.8
4-2	—	60	34	0	36.83	28.77	170.5	57.4
4-3	16	63	44	0.28	44.34	39.4	153.0	91.0
4-4	20	60	42	0.19	44.14	39.15	162.4	85.3
4-5	24	58	38	0.14	45.03	41.99	157.5	91.1
4-6	16	62	44	0.29	39.40	30.66	166.7	67.0
4-7	16	63	44	0.28	42.99	36.84	156.3	83.9
4-8	16	61	43	0.28	47.16	44.76	147.1	106.3
4-9	16	57	43	0.26	51.96	54.85	138.9	136.1
4-10	16	50	43	0.25	57.37	67.39	131.6	175.0
4-11	14	57	45	0.40	53.68	56.51	142.9	142.9
4-12	16	55	43	0.29	53.94	58.31	142.9	142.9
4-13	16	56	43	0.27	52.50	55.92	138.9	138.9
4-14	18	54	41	0.21	52.62	57.14	138.9	138.9
4-15	20	52	40	0.17	52.69	58.06	138.9	138.9
4-16	20	53	40	0.16	51.28	55.34	135.1	135.1
4-17	22	52	39	0.13	51.32	56.41	135.1	135.1

3.3 DESIGN EXAMPLE

In this chapter so far, we have left relatively little to the imagination of the designer. In practice the available prototypes are used to execute the designs in stripline or microstrip. In this section, a design from the previous paragraph is analyzed, using the computer analysis program SYMFOUR. Measured and calculated results for the example to be analyzed are also described in [4].

The analysis example is designated 3-14 in Table 3.3. The element values are 110.01 Ω for the end branch lines, or $y = 0.4545$; the series lines are unit elements of 46.33 Ω, which translates to $y = 1.0792$, while the center branch line has an admittance of 0.700 (71.43 Ω). The method of data entry into the computer program is shown in Appendix A, Section A2.4.3. The scattering parameters of the network are shown in Figure 3.3.

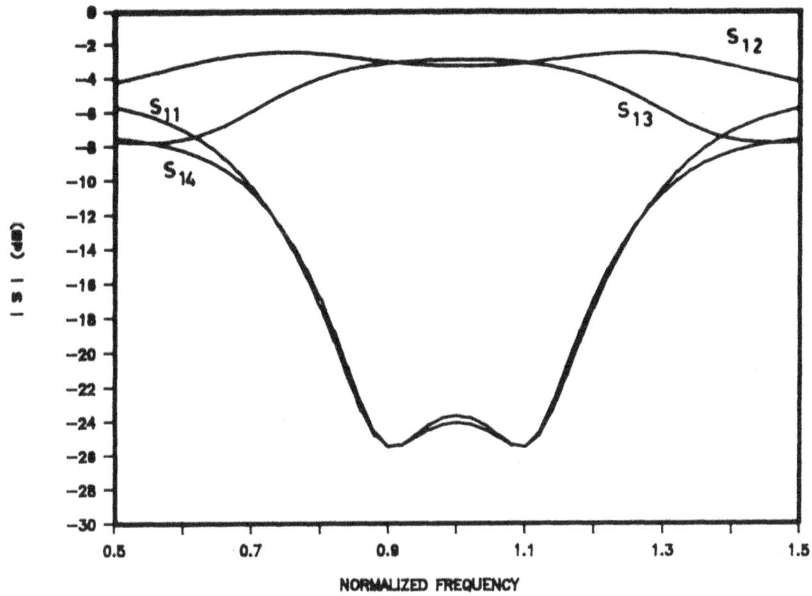

Figure 3.3 Calculated scattering parameters of third-order branch-line coupler.

3.4 WIDEBAND POWER DIVIDERS

The wideband power divider problem is very closely related to that of the wideband branch-line hybrid in a way very similar to the single-section circuits. The branch lines are not transmission lines, but rather resistors, and the two ports at one end are connected together to form the input port. Thus, in effect, the power divider is normally driven in the even mode, and the input reflection coefficient is also the even-mode reflection coefficient:

$$|\Gamma_1| = |\Gamma_e| \tag{3.32}$$

where the even and odd modes refer to output ports 2 and 3. For simplicity, the case of an even power split will be considered initially and extended later to include an uneven power split. For the former, referring to Figure 3.4, the reflection and transmission coefficients [5] are given by

$$|t_{12}| = |t_{13}| = \sqrt{\tfrac{1}{2}(1 - \Gamma_e^2)} \tag{3.33}$$

$$\Gamma_2 = \Gamma_3 = \tfrac{1}{2}(\Gamma_e + \Gamma_o) \tag{3.34}$$

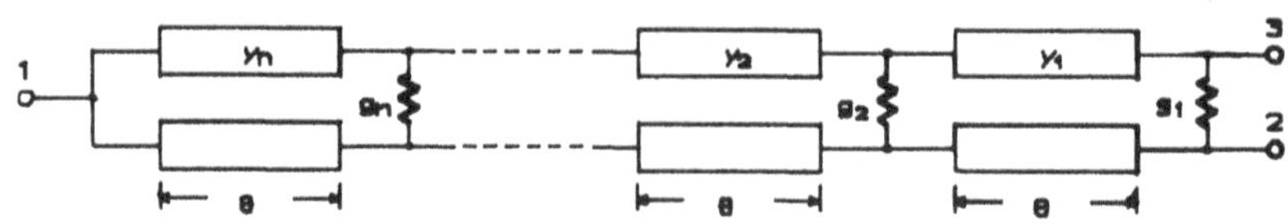

Figure 3.4 Multisection power divider.

$$t_{23} = \tfrac{1}{2}(T_e + T_o) \tag{3.35}$$

In the case of the branch-line coupler, there was no easy way in which the branch lines or the main lines could be coupled, because of the $\lambda/4$ separation between the lines. In the case of the power divider, the problem is reversed; under all circumstances, the resistors must be connected to the rest of the circuit by means of short lines, thereby implying that the physical separation between the lines cannot be large. This leads to two different types of networks: those with coupling between the lines and those without coupling. In the latter case, the even- and odd-mode admittances of the distributed part of the power divider are the same; in the case of coupled lines the odd- and even-mode admittances of course differ.

The odd- and even-mode networks that result from driving output ports 2 and 3 in the conventional way are shown in Figure 3.5. The resultant even-mode network is clearly a cascade of unit elements of admittance, $y_1, \ldots, y_n$, while the odd-mode network consists of the same unit elements but with conductances, $\tfrac{1}{2}g_1, \ldots, \tfrac{1}{2}g_n$, connected to ground as shown in Figure 3.5(b).

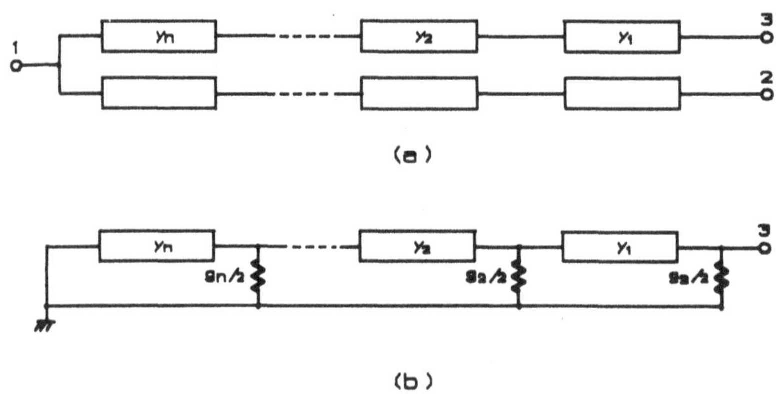

Figure 3.5 Odd- and even-mode networks of the general multisection power divider.

From Chapter 1, the cascade matrix for a unit element is

$$[A] = \frac{1}{\sqrt{1 - S^2}} \begin{bmatrix} 1 & Z_0 S \\ Y_0 S & 1 \end{bmatrix}$$

so that the cascade matrix of n unit elements for the even mode is given by

$$[A]_e = \left[\frac{1}{\sqrt{1 - S^2}}\right]^n \cdot \prod_{i=1}^{n} \begin{bmatrix} 1 & S/y_0 \\ y_0 S & 1 \end{bmatrix} \tag{3.36}$$

The cascade matrix for a shunt conductance is given by

$$[A] = \begin{bmatrix} 1 & 0 \\ g & 1 \end{bmatrix} \tag{3.37}$$

so that for the odd mode:

$$[A]_o = \left[\frac{1}{\sqrt{1 - S^2}}\right]^n \cdot \prod_{i=1}^{n} \begin{bmatrix} 1 & S/y_i \\ g_i + y_i S & S g_i/y_i + 1 \end{bmatrix} \tag{3.38}$$

We readily see that the driving point reflection coefficient in the conventional power divider application would be the same as that which would result in an even-mode application, a cascade of unit elements. Tables and methods for calculating prototype values for stepped transmission line transformers are available, and consequently, it is easy to specify the prototype by which to design the even-mode circuit [6]. However, the odd-mode circuit presents a problem.

The solution of the odd-mode circuit reduces to obtaining the conductances $g_1, \ldots, g_n$ such that the isolation and reflection coefficient at the two ports, 2 and 3, are optimum. Cohn [5] points out that it is much easier to optimize the odd-mode reflection coefficient and that in a number of computed cases it was found that if ρ_e and ρ_o are optimum, ρ_2 (and ρ_3) and t_{23} are very close to optimum.

3.4.1 Uncoupled Lines

The design procedures for the even-mode admittances for the two cases of uncoupled and coupled lines are the same, but the procedure for the odd-mode differs. We can show from eq. (3.38) that the reflection coefficient for the odd-mode network can be written as

$$\Gamma_o = \frac{P_n^n(S)}{P_n^d(S)} \tag{3.39}$$

where $P_n^n(S)$ and $P_n^d(S)$ are polynomials of order n in S.

3.4.1.1 Design Formulas for N = 2

For the case of $N = 2$, as shown in Figure 3.6, the expression for Γ_o is readily obtained. The input admittance is given by

$$y_{\text{in},o} = 2g_1 + y_1 \frac{(2g_2 + y_2/S) + Sy_1}{y_1 + (2g_2 + y_2/S)S} \tag{3.40}$$

and from this the odd-mode reflection coefficient follows as

$$\begin{aligned}\Gamma_o &= \frac{1 - y}{1 + y} \\ &= \frac{2g_2(1 - 2g_1) - y_1^2 - y_1y_2S^2 + [(y_1 + y_2)(1 - 2g_1) - 2g_2y_n]S}{2g_2(1 + 2g_1) + y_1^2 + y_1y_2S^2 + [(y_1 + y_2)(1 + 2g_1) + 2g_2y_n]S}\end{aligned} \tag{3.41}$$

The desired equiripple response of Γ_o and Γ_e is shown in Figure 3.7. The zeros of the numerator polynomial of Γ_o will determine the positions θ_3 and θ_4. All even factors of S are the real parts of the denominator, and odd factors are the imaginary part, and this yields two equations that have to be solved at θ_3:

$$2g_2(1 - 2g_1) - y_1^2 - y_1y_2S^2 = 0 \tag{3.42}$$

$$(y_1 + y_2)(1 - 2g_1) - 2g_2y_n = 0 \tag{3.43}$$

Solving for the conductances yields:

$$g_1 = \tfrac{1}{2}\sqrt{(y_1 + y_2)(y_1 + y_2S_3^2)} \tag{3.44}$$

$$g_2 = \tfrac{1}{2} - \frac{y_1g_2}{y_1 + y_2} \tag{3.45}$$

Now, we want to have a Chebyschev equiripple response, consequently, the reflection coefficient must be proportional to the second-order

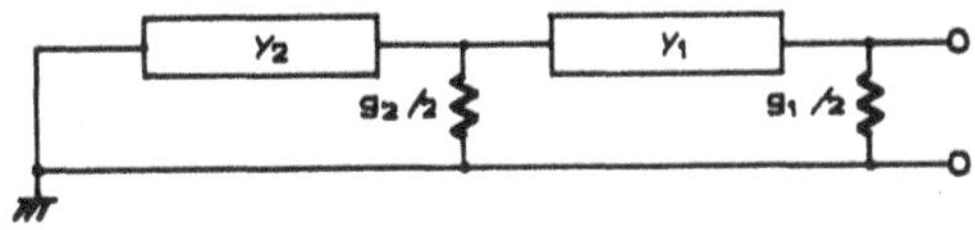

Figure 3.6 Odd-mode network for two-section power divider.

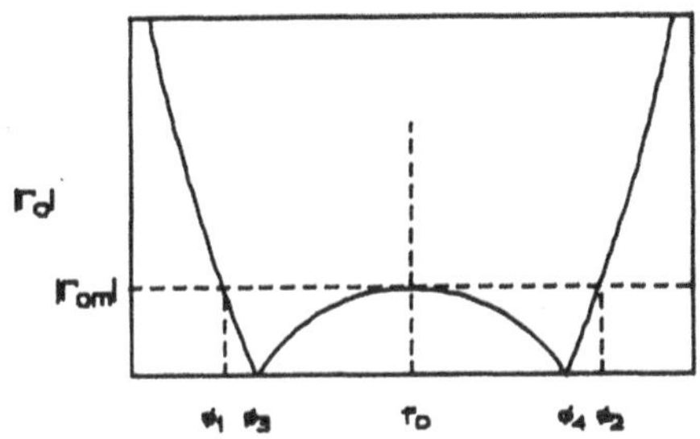

Figure 3.7 Ideal frequency response of reflection coefficient, $N = 2$.

Chebyschev polynomial:

$$|\Gamma_o| \propto T_2(x) = 2x^2 - 1 \tag{3.46}$$

The variable in this case is the normalized angular distance away from the center frequency, $x = S/S_c$, so that

$$x = \frac{\tan\theta}{\tan\theta_1} \tag{3.47}$$

However, the choice as to where exactly the zeros of the odd-mode reflection coefficient must lie is not known through synthesis, because the latter parameter is not being synthesized but both the sum and difference of the odd- and even-mode reflection coefficients. Consequently, (3.47) is not necessarily the optimal expression to use for x. In fact, Cohn [5], after evaluating a large number of designs, finds that the best expression to use is

$$x = \frac{\pi/2 - \phi}{\pi/2 - \phi_1} \tag{3.48}$$

so that the expression for relating ϕ_3 to ϕ_1 is given by

$$\phi_3 = \pi/2 - \frac{1}{\sqrt{2}}(\pi/2 - \phi_1)$$
$$= \pi/2 \left[1 - \frac{1}{\sqrt{2}}\left(\frac{f_2/f_1 - 1}{f_2/f_1 + 1}\right)\right] \tag{3.49}$$

Design Example

Lower band edge frequency, $f_1 = 0.7$
Upper band edge frequency, $f_2 = 1.3$
Relative bandwidth, $f_2/f_1 = 1.857$
Fractional bandwidth, $2(f_2/f_1 - 1)/(f_2/f_1 + 1) = 0.6$
Element values of Chebyschev transformer from [6]:

$$Z_1 = 1.21360 \quad Z_2 = 1.6480 \tag{3.50}$$
$$y_1 = 0.8240 \quad y_2 = 0.6068$$

Calculate

$\phi_3 = 1.2376 = 70.91^0$ [from (3.49)]

$g_2 = 0.5184$ [from (3.44)]

$g_1 = 0.2015$ [from (3.45)]

Figure 3.8 shows the layout of the power divider.

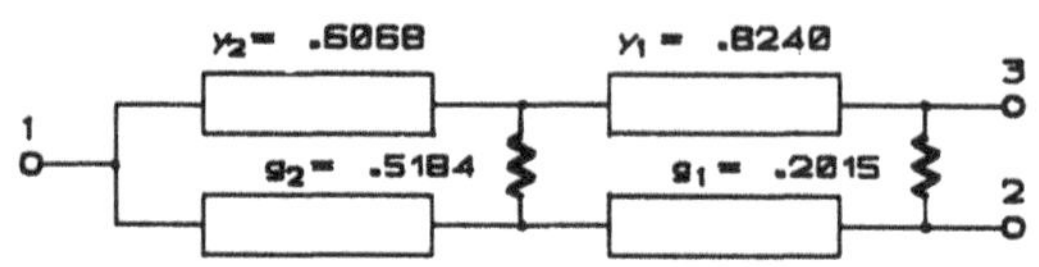

Figure 3.8 Example of a two-section power divider with 0.6 relative bandwidth.

Analysis

The calculated performance of the power divider is shown in Figure 3.9.

The isolation has a similar response and bandwidth. The return loss at the two output ports is more than 40 dB over the band. The power split is equal and flat.

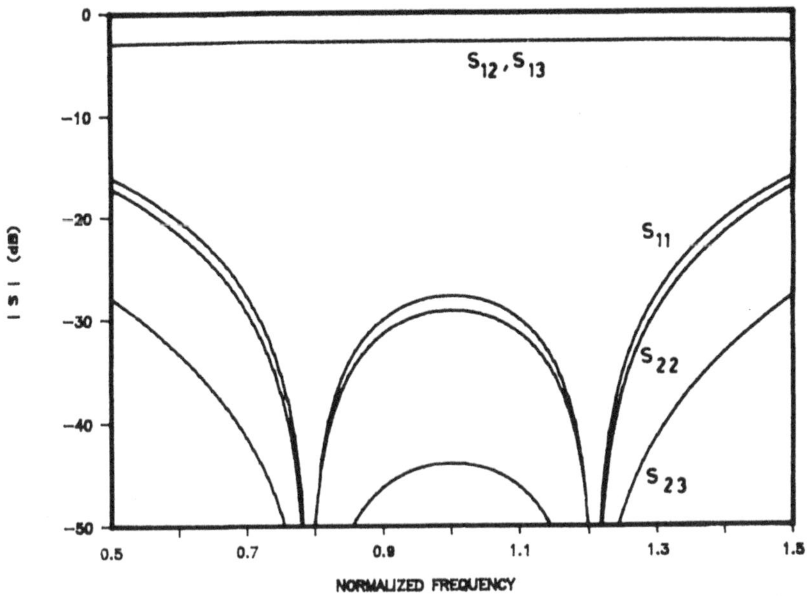

Figure 3.9 Calculated performance of two-section power divider with equi-ripple even-mode and odd-mode reflection coefficients.

3.4.1.2 General Design Formulas for $N \geqslant 3$

The procedure applied in the previous section for the case of $N = 2$ becomes too complex for higher order networks, and an approximate formula developed by Cohn [5] is used. He points out that, because a direct synthesis procedure is not available, other methods, including multiple reflections, also become too complex to manage, and a number of empirical equations that have been found to give good results are used.

As earlier, of course, the even-mode network is obtained from tables of Chebyschev transformers. The conductance values are calculated from

$$g_1 = 1 - y_1 \tag{3.51}$$

$$T_k = \frac{4y_{k-1}y_k}{(y_{k-1} + y_k + 2g_k)^2}, \quad k = 1 \text{ to } N \tag{3.52}$$

$$g_k = \frac{y_{k-1} - y_k}{y_{k-1}T_1T_2 \ldots T_{k-1}}, \quad k = 2 \text{ to } N - 1 \tag{3.53}$$

Finally, the last conductance is determined such that

$$y_{\text{in},o} = 1 + 0.7(S_{e,0} - 1)$$

where $S_{e,0}$ is the VSWR of the even-mode stepped transformer at band center. For a Chebyschev response, $S_{e,0} = 1$ for N odd, and $S_{e,0} = S_{em}$, the ripple value for N even. This is achieved by setting

$$g_N = \cfrac{\frac{1}{2}y_{N-1}^2}{-2g_{N-1} + \cfrac{y_{N-2}^2}{-2g_{N-2} + \cfrac{y_{N-3}^2}{\cfrac{\cdots}{-2g_2 + \cfrac{y_1^2}{-2g_1 + 1 + 0.7(S_{e,0} - 1)}}}}} \tag{3.54}$$

The maximum VSWR is given by

$$S_{1m} = S_{em} \tag{3.55}$$

$$S_{2m} = S_{3m} \approx 1 + 0.2(S_{em} - 1) \tag{3.56}$$

while the minimum isolation is given approximately by

$$I_m \approx 20 \log\left(\frac{2.35}{S_{em} - 1}\right) \text{ dB} \tag{3.57}$$

Design Example

Lower band edge frequency, $f_1 = 0.3$
Upper band edge frequency, $f_2 = 1.7$
Relative bandwidth, $f_2/f_1 = 5.6667$
Fractional bandwidth, $2(f_2/f_1 - 1)/(f_2/f_1 + 1) = 1.4$
Element values of Chebyschev transformer from [6],

$$z_1 = 1.165 \qquad y_1 = 0.8584$$
$$z_2 = 1.320 \qquad y_2 = 0.7576$$
$$z_3 = 1.515 \qquad y_3 = 0.6600$$
$$z_4 = 1.717 \qquad y_4 = 0.5852$$

Application of eqs. (3.51)–(3.54) yields

$$g_1 = 0.1416$$
$$g_2 = 0.1569$$
$$g_3 = 0.2431$$
$$g_4 = 0.3718$$

The element values of the power divider are shown in Figure 3.10. The following performance prediction follows from eqs. (3.55) and (3.56):

$$S_{11} = -20.24 \text{ dB}$$
$$S_{21} = S_{31} = -3 \text{ dB}$$
$$S_{22} = S_{33} \leq -33.5 \text{ dB}$$
$$S_{23} = S_{32} \leq -20.75 \text{ dB}$$

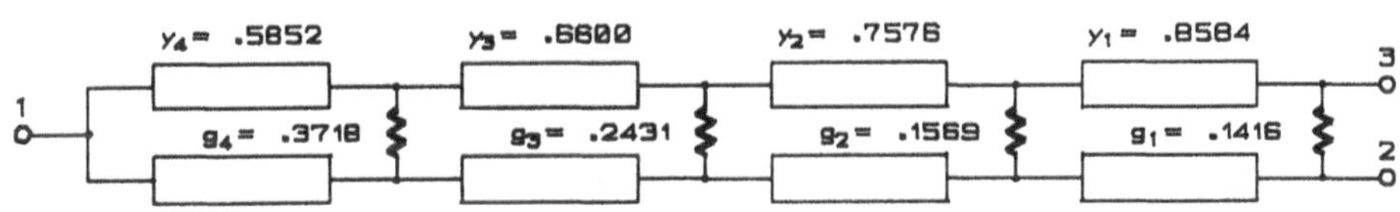

Figure 3.10 Element values of four-section power divider with 140% relative bandwidth.

Analysis

The even- and odd-mode networks were programmed into SYMTHREE and the various network parameters calculated. Figure 3.11 shows the performance of the power divider. The input reflection coefficient is exactly as had been designed, and the reflection coefficient and isolation

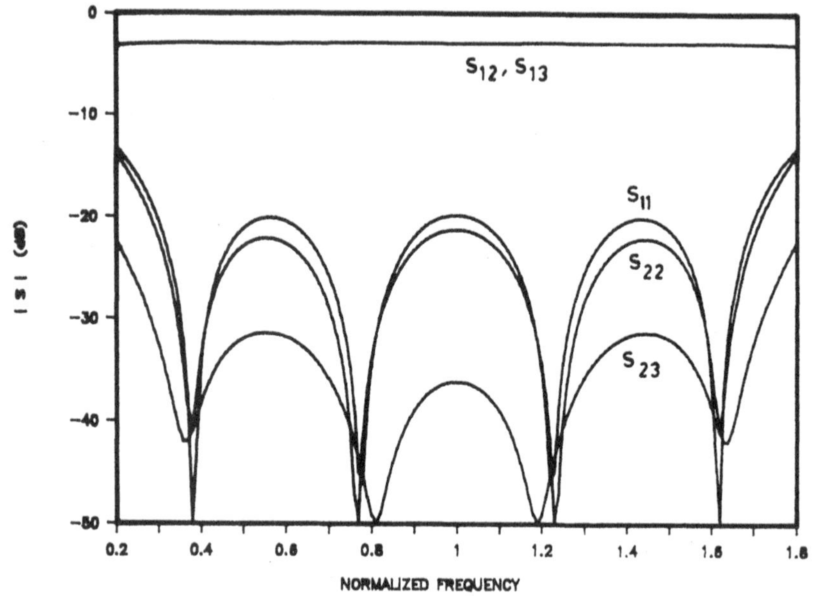

Figure 3.11 Calculated performance of four-section power divider.

had been predicted adequately by the approximate theory. Finally, it is of interest to note that Cohn [5] considers the limit of useful bandwidth to be 2.5 : 1 for two sections, 4 : 1 for three sections, 5.5 : 1 for four sections, and 10 : 1 for ten sections.

3.4.2 Power Dividers with Coupled Lines

If the pairs of lines of a power divider are allowed to move closer together, so that the fields around the lines become coupled, the even- and odd-mode admittances of the resulting four ports differ. In cases of tight coupling, there are considerable differences between the even and odd modes, and the design of a power splitter cannot be executed by the method described in the previous section.

The even-mode network is still essentially the same as before and can be designed as such. In the case of the odd-mode network, though, the line admittances are now unknowns as well, and the previously difficult synthesis problem now becomes completely intractable. Ekinge [7] describes a method where the odd-mode network has been optimized by means of a computer program [8]. Unfortunately, tabulated element values are not available.

In Nyström [9] the network is modified to include only one isolation resistor, as shown in Figure 3.12. The equiripple function that was used in Section 3.2.2.2 is used in this case:

$$P_L = 1 + \frac{kP_n^2(x)}{(1 - x^2)} \tag{3.58}$$

and the nth-order Chebyschev polynomial is given by

$$P_n(x) = \tfrac{1}{2}(1 + \sqrt{1 - x_c^2})T_n\left(\frac{x}{x_c}\right) - \tfrac{1}{2}(1 - \sqrt{1 - x_c^2})T_{n-2}\left(\frac{x}{x_c}\right) \tag{3.59}$$

where k is a constant and

$$x = \cos\,(\pi/2\; f/f_0), \qquad x_c = \cos\,(\pi/2\; f_c/f_0) \tag{3.60}$$

The even-mode network is once again obtained directly from tables, and the odd-mode network synthesized as described in Section 3.2. As earlier, it is much simpler to make use of tabulated values, and Table 3.5 shows coupled-line impedance values for a number of coupled-line power dividers from [9] for the case $N = 2$. Values for $N = 3$ to 6 are tabulated in Tables 3.6 to 3.9.

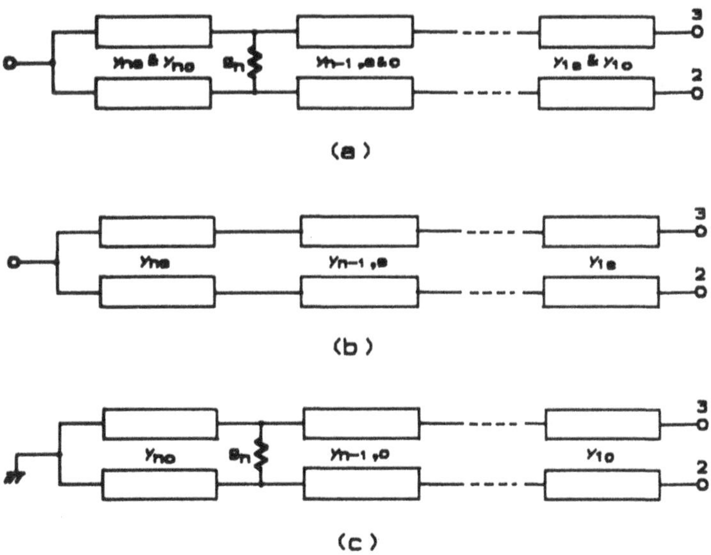

Figure 3.12 Broadband hybrid with only one isolation resistor, (a). The even-mode network is shown in (b) and the odd-mode network in (c).

Table 3.5 Element Values for −3 dB Hybrids (N = 2)
(© 1981 IEEE, reprinted with permission, from [9])

B	VSWR	n	z_{0e}	z_{0o}	g_n^{-1}
1.61	1.05	1	1.204	0.765	—
		2	1.661	1.133	1.200
1.96	1.1	1	1.218	0.748	—
		2	1.642	1.197	1.231
2.60	1.2	1	1.244	0.734	—
		2	1.607	1.345	1.294

Table 3.6 Element Values for −3 dB Hybrids (N = 3)
(© 1981 IEEE, reprinted with permission, from [9])

B	VSWR	n	z_{0e}	z_{0o}	g_n^{-1}
2.34	1.05	1	1.125	0.844	—
		2	1.414	0.551	—
		3	1.778	1.037	0.853
2.94	1.10	1	1.147	0.825	—
		2	1.414	0.555	—
		3	1.743	1.143	0.906
3.97	1.20	1	1.186	0.797	—
		2	1.414	0.563	—
		3	1.686	1.360	1.000

Table 3.7 Element Values for −3 dB Hybrids (N = 4)
(© 1981 IEEE, reprinted with permission, from [9])

B	VSWR	n	z_{0e}	z_{0o}	g_n^{-1}
3.15	1.05	1	1.090	0.888	—
		2	1.280	0.667	—
		3	1.562	0.462	—
		4	1.853	1.007	0.674
4.01	1.1	1	1.116	0.865	—
		2	1.296	0.865	—
		3	1.543	0.445	—
		4	1.792	1.171	0.746
5.44	1.2	1	1.159	0.831	—
		2	1.317	0.653	—
		3	1.518	0.471	—
		4	1.725	1.474	0.860

Table 3.8 Element Values for −3 dB Hybrids (N = 5)
(© 1981 IEEE, reprinted with permission, from [9])

B	VSWR	n	z_{0e}	z_{0o}	g_n^{-1}
4.01	1.05	1	1.072	0.912	—
		2	1.207	0.745	—
		3	1.414	0.536	—
		4	1.657	0.356	—
		5	1.886	1.024	0.574
5.1	1.1	1	1.099	0.887	—
		2	1.230	0.739	—
		3	1.414	0.544	—
		4	1.626	0.378	—
		5	1.820	1.228	0.652
6.94	1.2	1	1.144	0.849	—
		2	1.236	0.709	—
		3	1.414	0.556	—
		4	1.584	0.415	—
		5	1.748	1.606	0.777

Table 3.9 Element Values for −3 dB Hybrids (N = 6)
(© 1981 IEEE, reprinted with permission, from [9])

B	VSWR	n	z_{0e}	z_{0o}	g_n^{-1}
4.92	1.05	1	1.061	0.927	—
		2	1.163	0.799	—
		3	1.319	0.626	—
		4	1.517	0.452	—
		5	1.720	0.310	—
		6	1.885	1.098	0.521
6.23	1.1	1	1.089	0.900	—
		2	1.190	0.774	—
		3	1.331	0.620	—
		4	1.502	0.466	—
		5	1.681	0.336	—
		6	1.837	1.034	0.593
8.54	1.2	1	1.135	0.861	—
		2	1.229	0.747	—
		3	1.384	0.618	—
		4	1.484	0.491	—
		5	1.627	0.378	—
		6	1.762	1.754	0.723

Above $N = 3$, the coupling of the second to the last section becomes extremely tight, as we will see later.

3.4.3 Coupled-Line Equations

In this section, we will consider briefly some aspects of coupled-line realization. As Chapter 4 is devoted to coupled-line realization, the reader is referred there for further and more detailed information. In order to avoid confusion, the conventional design equations will not be used here. (The possible confusion revolves around the use of the expression *Characteristic Impedance:* In directional couplers with coupled lines, all sections will have a coupled-line characteristic impedance equal to the system characteristic impedance Z_0. In coupled-line power splitters, the characteristic impedance Z_0 refers only to the system characteristic impedance.)

From eqs. (1.22)–(1.25) it follows quite logically that the odd and even capacitances for coupled lines can be obtained from

$$C_{0e}/\epsilon = \frac{\eta}{\sqrt{\epsilon_r}} \cdot \frac{1}{Z_{0e}} \tag{3.61}$$

$$C_{0o}/\epsilon = \frac{\eta}{\sqrt{\epsilon_r}} \cdot \frac{1}{Z_{0o}} \tag{3.62}$$

$$\Delta C = \tfrac{1}{2}\,(C_{0o} - C_{0e}) \tag{3.63}$$

With these equations, the cross section of any type of physical realization of the coupled lines can readily by found, as we will illustrate by the example to follow. The mutual capacitance values for the tightest section for the cases tabulated earlier are given in Table 3.10, as calculated from (3.63) for a general case of $\sqrt{\epsilon_r} = 1.5$. For stripline realization, it will be found that only the cases of $N = 2$ and 3 are realizable with simple side-coupled stripline. If offset parallel-coupled lines are used, the connection of the isolation resistor, which will always be in the overcoupled region, becomes complicated, as a connection has to be made through the center board.

Table 3.10 **Maximum Mutual Capacitance**

N	B	VSWR	ΔC_m
	1.61	1.05	1.198
2	1.96	1.1	1.297
	2.60	1.2	1.404
	2.34	1.05	2.784
3	2.94	1.1	2.751
	3.79	1.2	2.687
	3.15	1.05	3.831
4	4.01	1.1	4.019
	5.44	1.2	3.680
	4.01	1.05	5.543
5	5.01	1.1	5.103
	6.94	1.2	4.470
	4.92	1.05	6.646
6	6.23	1.1	5.985
	8.54	1.2	5.104

3.5 DESIGN EXAMPLE

As an example, consider the design of a three-section power divider, with a bandwidth ratio of 2.94 or relative bandwidth of 98 percent. The element values for this design are given from Table 3.6 as provided in Table 3.11.

Table 3.11 **Element Values for Design Example**

B	VSWR	n	z_{0e}	z_{0o}	g_n^{-1}
2.94	1.10	1	1.147	0.825	—
		2	1.414	0.555	—
		3	1.743	1.143	0.906

From eq. (3.63) calculate

$$\Delta C_1 = 0.855$$

$$\Delta C_1 = 2.326$$

$$\Delta C_1 = 0.757$$

The even- and odd-mode impedances are calculated, as is the input to LDIMS. The dimensions are given for a cross section as shown in Figure 3.13(a). (See Appendix B for operation of the computer program.)

Material properties,	$\epsilon_r = 2.4$;
Ground plane spacing,	$b = 3.3$ mm
Center strip,	$s = 0.125$ mm
Center-strip ratio,	$s/b = 0.3788$

Section 1

$Z_{0e} = 57.35\ \Omega$	$Z_{0o} = 41.25$
$w = 2.516$ mm	$w_c = -0.581$ mm

Section 2

$Z_{0e} = 70.70\ \Omega$	$Z_{0o} = 27.75\ \Omega$
$w = 2.117$ mm	$w_c = 0.048$ mm

Section 3

$Z_{0e} = 87.15\ \Omega$	$Z_{0o} = 57.15\ \Omega$
$w = 1.285$ mm	$w_c = -0.683$ mm

Figure 3.13(b) shows the layout of the power splitter. The performance of the network was also analyzed by means of SYMTHREE, which is shown in Figure 3.14.

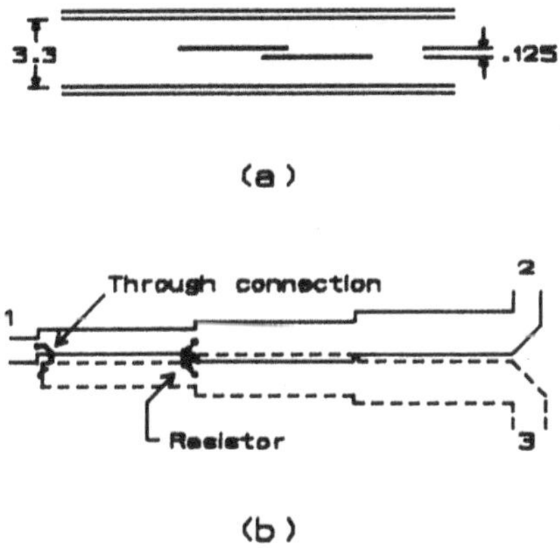

Figure 3.13 Cross section of offset parallel coupled-line power splitter, (a). The artwork pattern is shown in (b).

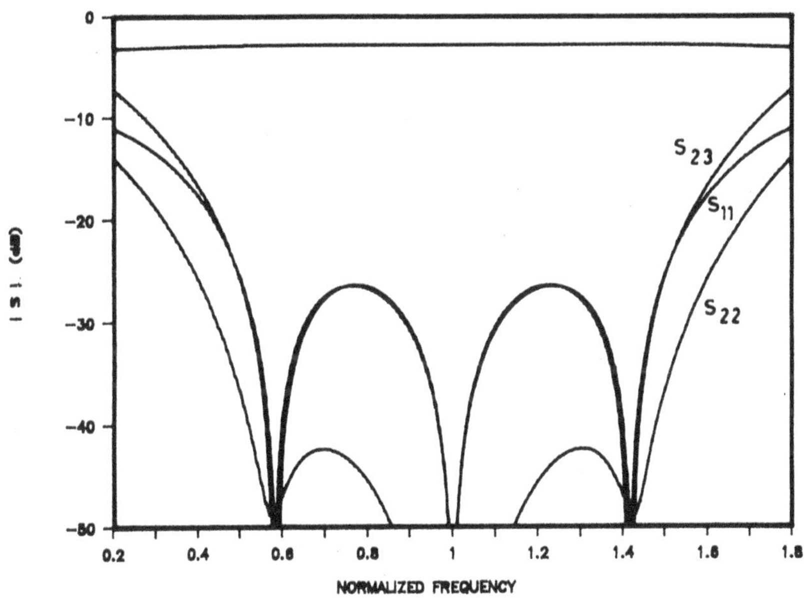

Figure 3.14 Frequency response of coupled-line power splitter.

REFERENCES

[1] R. Levy and L. F. Lind, "Synthesis of Symmetrical Branch-Guide Directional Couplers," *IEEE Trans. Microwave Theory Tech.*, Vol. MTT-16, February 1968, pp. 80–89.

[2] J. A. G. Malherbe, *op. cit., Microwave Transmission Line Filters,* Artech House, Dedham, MA, 1979.

[3] R. Levy, "Zolotarev Branch-Guide Couplers," *IEEE Trans. Microwave Theory Tech.*, Vol. MTT-21, February 1973, pp. 95–99.

[4] M. Maraguchi, T. Yukitake, and Y. Naito, "Optimum Design of 3-dB Branch-Line Couplers Using Microstrip Lines," *IEEE Trans. Microwave Theory Tech.*, Vol. MTT-31, August 1983, pp. 674–678.

[5] S. B. Cohn, "A Class of Broadband Three-Port TEM-Mode Hybrids," *IEEE Trans. Microwave Theory Tech.*, Vol. MTT-16, February 1968, pp. 110–116.

[6] G. L. Matthaei, L. Young, and E. M. T. Jones, *Microwave Filters, Impedance-Matching Networks and Coupling Structures,* Artech House, Dedham, MA, 1980.

[7] R. B. Ekinge, "A New Method of Synthesizing Matched BroadBand TEM-Mode Three-Ports," *IEEE Trans. Microwave Theory Tech.*, Vol. MTT-19, January 1971, pp. 81–87.

[8] R. B. Ekinge, "Hybrid-T Synthesis," *IEEE Trans. Microwave Theory Tech.*, Vol. MTT-18, November 1970, pp. 1006–1007.

[9] G. L. Nyström, "Synthesis of Broad-Band 3-dB Hybrids Based on the Two-Way Power Divider," *IEEE Trans. Microwave Theory Tech.*, Vol. MTT-29, March 1981, pp. 189–194.

Chapter 4
Single-Section Coupled-Line Directional Couplers

4.1 INTRODUCTION

In this chapter we describe the design and realization of directional couplers that make use of the coupled lines. In Chapter 1 the general network model was used to describe the network properties of a general coupled-line structure. In this chapter, we will use synthesis procedures to yield element values and circuit properties that can be fitted onto the network model to obtain the physical network with the desired performance. The accent will be on the cross-sectional property, as well as the properties of a single section that can be used later as an element in a multisection coupler.

4.2 COUPLER ACTION

4.2.1 Responses

Figure 4.1(a) shows the basic coupler that is to be realized. Physically, the coupler consists of a pair of coupled lines, shown schematically in 4.1(b). Choosing the convention of an isolated port 3, we can see that, if the network is to operate as a directional coupler, there can be no voltage across terminal 3 (at any frequency). Referring to the network model in Figure 4.1(c), this means that the unit element Z_2 is open circuited and consequently completely disconnected from the rest of the circuit at its right hand side. From its input side on the left, it now looks like a Richards' capacitor, and the equivalent circuit shown in Figure 4.1(d) can be drawn.

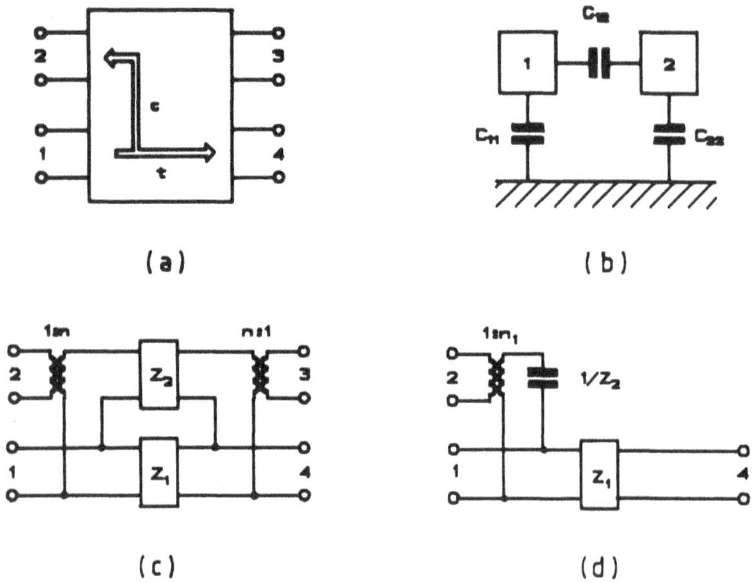

Figure 4.1 **Basic directional coupler (a), with schematic representation of physical coupler (b). The network model of the coupler is shown in (c), with the equivalent circuit shown in (d) [1].**

For a transmission line of length l, let

β = the propagation constant on the line;
Z_0 = the characteristic impedance of the line;
Z_L = the load impedance;
V_G = the generator voltage;
V_L = the voltage at the load;
$y_L = 1/z_L = Z_o/Z_L$ = the normalized load admittance;
$\theta = \beta l$ = the electrical length of the line.

From the basic transmission line theory (1.14):

$$V_G = V_L(\cos\theta + jy_L \sin\theta) \tag{4.1}$$

Rewriting in terms of the variables in Figure 4.1(d) and normalizing the characteristic impedance of line Z_1 to the port impedance Z_0, as z_1:

$$\frac{V_4}{V_1} = \frac{1}{\cos\theta + jz_1 \sin\theta}$$

$$= \frac{y_1}{y_1 \cos\theta + j \sin\theta} \tag{4.2}$$

The input impedance to the capacitor formed by Z_2 is given by

$$z_c = -jz_2 \cot\beta l \tag{4.3}$$

The voltage at the coupled port can be obtained by noting that the input voltage is divided between the series capacitor and the load at port 2 through the transformer:

$$\frac{V_2}{V_1} = \frac{1}{n} \cdot \frac{n^2}{n^2 - jz_2 \cot\theta}$$

$$= \frac{j1/n \sin\theta}{z_2/n^2 \cos\theta + j \sin\theta} \tag{4.4}$$

Application of the principle of conservation of energy means that the power into port 1 must go to ports 2 and 4 only;

$$\left|\frac{V_4}{V_1}\right|^2 + \left|\frac{V_2}{V_1}\right|^2 = 1 \tag{4.5}$$

Substitution of the expressions for the voltage ratios at midband, where $\theta = \pi/2$, from eqs. (4.2) and (4.4) in (4.5) gives the condition:

$$1/n^2 + y_1^2 = 1$$

or

$$y_1^2 = 1 - 1/n^2 = 1 - c^2 \tag{4.6}$$

where

$$c = 1/n \tag{4.7}$$

The condition for equal width lines (1.27) states that

$$z_2 = z_1(n^2 - 1) \qquad \text{or} \qquad y_1 = y_2(n^2 - 1) \tag{4.8}$$

and if eqs. (4.6), (4.7), and (4.8) are substituted into (4.2) and (4.4), the voltage ratios for the coupler can be expressed as

$$\frac{V_4}{V_1} = \frac{\sqrt{(1 - c^2)}}{\sqrt{(1 - c^2)}\cos\theta + j\sin\theta} \tag{4.9}$$

$$\frac{V_2}{V_1} = \frac{jc\sin\theta}{\sqrt{(1 - c^2)}\cos\theta + j\sin\theta}\,. \tag{4.10}$$

From (4.9) and (4.10) it can be seen that, irrespective of frequency, there is a 90° phase shift between the outputs at ports 2 and 4.

4.2.2 Isolation

Note that if the expressions (4.9) and (4.10), which are frequency dependent, are substituted back into (4.5), the power conservation condition is still met. This implies the very important result that the isolated port remains isolated at all frequencies.

4.2.3 Match

Further insight is obtained by studying the input impedance at each of the ports. The input impedance at port 1 is the shunt of the coupling branch to port 2 and the through branch to port 4:

$$\begin{aligned} y_{\text{in}} &= y_1 \frac{1 + jy_1 \tan\theta}{y_1 + j\tan\theta} + \frac{1}{n^2 - jz_2 \cot\theta} \\ &= 1 \end{aligned}$$

Similarly, all the other ports can be shown to be matched, irrespective of frequency.

4.2.4 Coupler Condition

The directional coupler condition (4.6), when combined with the condition for equal width lines (4.8), and the equations for the odd- and even-mode impedance from (1.28) and (1.29) gives the important result

$$z_{0e}z_{0o} = 1 \tag{4.11}$$

and in unnormalized form:

$$Z_{0e}Z_{0o} = Z_0^2 \tag{4.12}$$

4.2.5 Coupling

A pair of coupled lines has three variables: the two line widths and the gap. In the coupler application, however, only two of these are independent, so that it is possible to specify the coupler completely through the odd- and even-mode impedances. Indeed, in many cases the available design prototypes are specified in terms of the odd and even modes.

On the other hand, the same information is contained in the characteristic impedance and the mutual coupling, and the latter variable can be used as a direct indication of the realizability of a section by means of a given physical construction. In terms of the even and odd modes, from (1.24), (1.26), and (1.28), follows

$$n = 1 + \frac{1/Z_{0e}}{\frac{1}{2}(1/Z_{0o} - 1/Z_{0e})}$$
$$c = \frac{Z_{0e} - Z_{0o}}{Z_{0e} + Z_{0o}} \tag{4.13}$$

Substituting (4.12) into (4.13) gives

$$Z_{0e} = \sqrt{\frac{1 + c}{1 - c}}\, Z_0 \tag{4.14}$$

or

$$z_{0e} = \sqrt{\frac{1 + c}{1 - c}} \tag{4.15}$$

In terms of the mutual capacitance

$$\Delta C = \tfrac{1}{2}(C_{0o} - C_{0e})$$

$$\Delta C = \frac{\eta}{Z_0\sqrt{\epsilon_r}} \cdot \frac{c}{\sqrt{1 - c^2}} \tag{4.16}$$

Example

The following example illustrates the procedure of obtaining the capacitance matrix and the typical variation of S-parameters *versus* frequency.

Characteristic impedance,	Z_0	$= 50\ \Omega$
Coupling coefficient,	c	$= .7071$
Coupling coefficient (dB),		$= -3.01$
From (4.14),	Z_{0e}	$= 120.71\ \Omega$
From (4.12),	Z_{0o}	$= 20.71\ \Omega$

Note that, at this stage, the mutual capacitance cannot be calculated as it is a function of the medium in which the coupler is to be realized. The S-parameter responses are shown for this example in Figure 4.2; the computer screen display for running the program is shown in Appendix A, Figure A7. Only the coupling and transmission are shown.

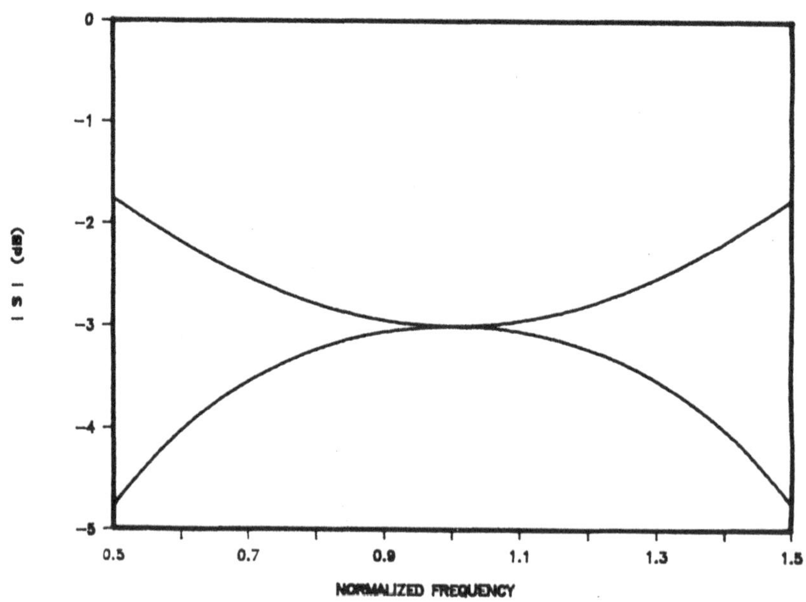

Figure 4.2 Computer screen display and calculated responses of S-parameters *versus* frequency for the design example of a -3 dB coupled-line coupler.

The reflection coefficient and isolation were constant at a level of −75 dB. This is apparently contradictory to the theoretical result, but in fact it is not. The nonideal response of the coupler for match and isolation is because the figures in the analysis had been rounded to three decimals and the analysis is extremely sensitive to rounding at such low power levels.

4.3 COUPLERS WITH UNEQUAL CHARACTERISTIC IMPEDANCES

There are cases where the lines of a coupled-line coupler are not of equal width; in this case, the ports are not all terminated in the same characteristic impedance. Figure 4.3 shows a pair of coupled lines with unequal characteristic impedances that function as a directional coupler between the levels of Z_0 and zZ_0 Ω.

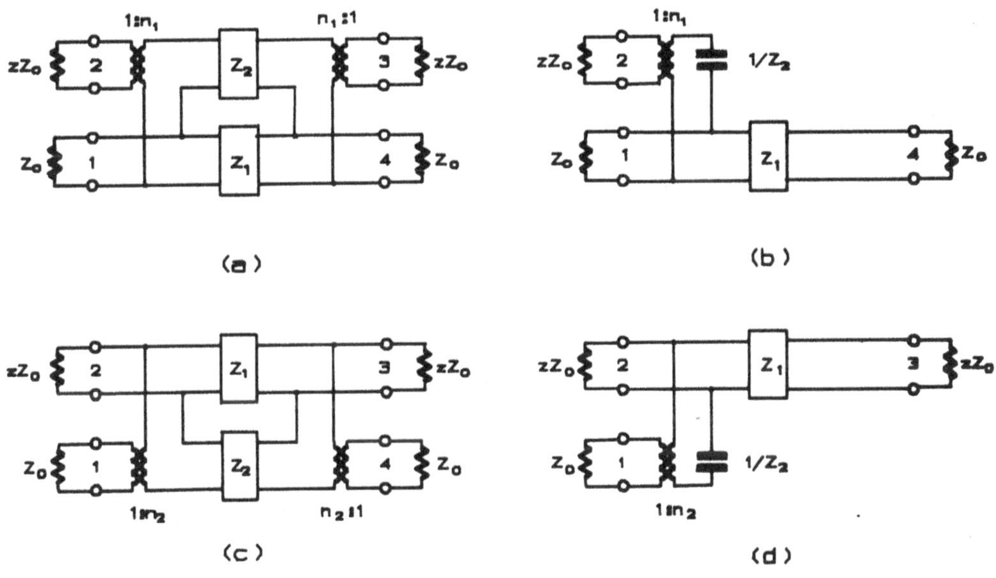

Figure 4.3 **Directional coupler with unequal characteristic impedances, (a). The circuit simplifies to that at (b) if no power flows in port 3. The same coupler is modeled in (c) with the ports reversed. It is simplified in (d).**

The solution of the problem is obtained by applying the process performed in the previous section twice; that is, a set of equations are obtained from Figure 4.3(a) and then the equivalent circuit is drawn for the network with the port roles reversed as in Figure 4.3(c). It is not necessary to repeat the complete derivation of all the coupling properties, as the frequency responses are similar to those for the conventional coupler. Referring to Figure 4.3(b), the midband coupling can be written down by inspection as

$$k = |S_{21}| = \frac{1}{n_1\sqrt{z}} \tag{4.17}$$

$$n_1 = 1/(k\sqrt{z}) \tag{4.18}$$

From conservation of power, with unit voltage applied to port 1, follows

$$y_1^2 + \frac{1}{n_1^2 z} = 1$$

$$y_1 = \sqrt{1 - k^2} \tag{4.19}$$

From (1.19) and (1.20) follow expressions for the capacitances in the equivalent circuit of Figure 4.3(b):

$$C_{11}/\epsilon = \frac{\eta}{Z_0\sqrt{\epsilon_r}}[y_1 - y_2(n_1 - 1)] \tag{4.20}$$

$$C_{12}/\epsilon = \frac{\eta}{Z_0\sqrt{\epsilon_r}} y_2 n_1 \tag{4.21}$$

The same procedure is now repeated for the equivalent circuit of Figure 4.3(d):

$$k = |S_{21}| = \frac{\sqrt{z}}{n_2} \tag{4.22}$$

$$n_2 = \sqrt{z}/k \tag{4.23}$$

From conservation of power, with unit voltage applied to port 2, follows

$$y_1^2 + \frac{1}{n_2^2 z} = \frac{1}{z^2}$$

$$y_1 = \sqrt{1 - k^2/z} \tag{4.24}$$

From the definition of n_2 follows

$$n_2 = 1 + \frac{C_{11}/\epsilon}{C_{12}/\epsilon}$$

$$C_{11}/\epsilon = C_{12}/\epsilon(n_2 - 1) \tag{4.25}$$

Eliminating C_{11}, y_1, and y_2 from eqs. (4.19), (4.20), (4.21), and (4.25) with substitution of the values of n_1 and n_2 from (4.18) and (4.23) gives

$$C_{12}/\epsilon = \frac{\eta}{Z_0\sqrt{\epsilon_r}} \cdot \frac{1}{\sqrt{z}} \cdot \frac{k}{\sqrt{1 - k^2}} \tag{4.26}$$

and

$$C_{11}/\epsilon = C_{12}/\epsilon \left(1 - \frac{1}{k\sqrt{z}}\right) \tag{4.27}$$

$$C_{22}/\epsilon = C_{12}/\epsilon \left(1 - \frac{\sqrt{z}}{k}\right) \tag{4.28}$$

These are design equations for the case where the characteristic impedances are not equal. Although their derivation is different, the result agrees with that obtained in [2].

4.4 MULTIPLE COUPLED LINES

In many instances, a specific type of physical construction is particularly desirable. For instance, it would not be acceptable to make use of thick coupled metal bars between parallel plates for a coupler to be integrated into a lightweight microstrip circuit. In such cases, the coupling between lines can be increased by making use of a number of lines alternately strapped together to form two coupled structures. This is shown schematically in Figure 4.4 for the case of n even.

For the set of lines connected together as a, the total self-capacitance is given by

$$C_a/\epsilon = \sum_i C_{ii}, \qquad i = 1, 3, 5, \ldots \tag{4.29}$$

while for the lines b

$$C_b/\epsilon = \sum_i C_{ii}, \qquad i = 2, 4, 6, \ldots \tag{4.30}$$

and for the mutual capacitance

$$C_{ab} = \sum_{i=1}^{n-1} C_{ij} \tag{4.31}$$

These capacitances can now be used in eqs. (4.14)–(4.16) or (4.26)–(4.28) in the place of the normal two-line cross-sectional capacitances. It is clear that, as the number of lines are increased, the mutual capacitance can be made equal to the number of gaps times the capacitance of a single gap.

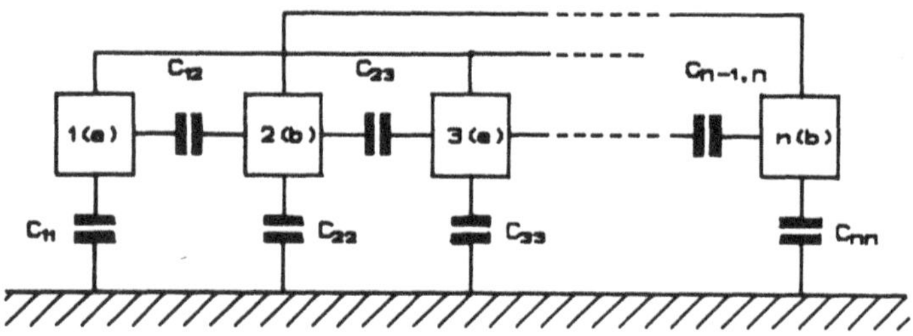

Figure 4.4 Schematic cross section of multiple coupled lines.

4.4.1 The Lange Coupler

The use of four lines in the specific configuration, shown in Figure 4.5, was first proposed by Lange [3] and affords tight coupling. Due to the specific problems associated with the design of four or more coupled microstrip lines, the design of this type of coupler is treated separately. In theory, the performance of the Lange coupler and its "unfolded" version is the same. In practice, the unfolded version uses fewer bonding wires and affords the designer the option of inverting the positions of the through and isolated ports. This might be an advantage in certain physical configurations.

If the mutual capacitance between the n lines (n is an even number) is divided equally, the total mutual capacitance is given by

$$C_{ab} = (n - 1)C_{12} \tag{4.32}$$

The lines are all made to be the same width, and all the gaps are equal. Under these circumstances, to good approximation:

$$C_{22} = C_{11} - \frac{C_{11}C_{12}}{C_{11} + C_{12}} \tag{4.33}$$

$$C_a = C_{11} + (\tfrac{1}{2}n - 1)C_{22}$$

$$= C_{11} + (\tfrac{1}{2}n - 1)\left(C_{11} - \frac{C_{11}C_{12}}{C_{11} + C_{12}}\right) \tag{4.34}$$

Setting

$$y_{0e} = C_{11}/\epsilon \cdot \sqrt{\epsilon_r}/\eta \tag{4.35}$$

$$y_{0o} = (C_{11}/\epsilon + C_{12}/\epsilon) \cdot \sqrt{\epsilon_r}/\eta \tag{4.36}$$

Substituting (4.26) and (4.27) into (4.33)–(4.36), after considerable manipulation as per [5] and [6], gives

$$Z_{0o} = Z_0 \left(\frac{1 - c}{1 + c}\right)^{1/2} \cdot \frac{(n - 1)(1 + q)}{(c + q) + (n - 1)(1 - c)} \tag{4.37}$$

$$Z_{0e} = Z_{0o} \frac{(c + q)}{(n - 1)(1 - c)} \tag{4.38}$$

where

$$q = [c^2 + (1 - c^2)(n - 1)]^2 \tag{4.39}$$

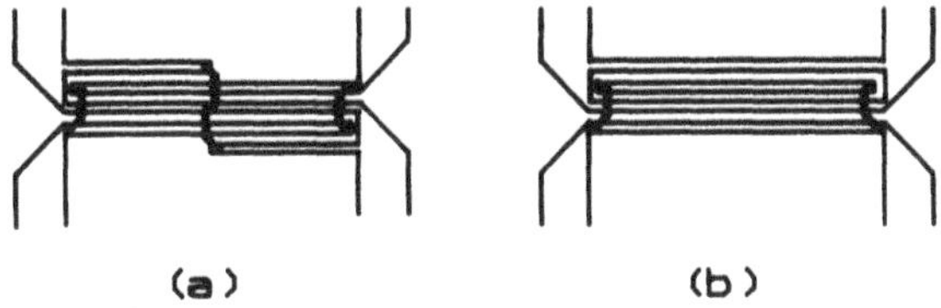

Figure 4.5 Lange coupler. The original coupler is shown in (a), with the so-called "unfolded" version [4] shown in (b).

The method whereby the even- and odd-mode impedances are converted to coupled-line dimensions will be illustrated later in an example. Note that, if $n = 2$, the conventional expressions for coupling and characteristic impedance results.

4.5 DESIGN EXAMPLES

The design of single-section coupled-line directional couplers will be treated by studying a number of design examples, incorporating loose and tight coupling, as well as a number of different types of physical constructions. The reader is referred to Appendix B for information on the properties of a number of popular media for coupler realization.

4.5.1 Stripline Couplers, Low Coupling

The design equations and definition of the cross section are given in Section B3.1, together with an example of the calculations of dimensions using the computer program LDIMS. As discussed in B3.3, the maximum coupling realizable with conventional side-coupled striplines is somewhere between $\Delta C = 1.2$ and $\Delta C = 1.6$. From (4.16):

$$c^2 = \frac{(\Delta C Z_0 \sqrt{\epsilon_r}/\eta)^2}{1 + (\Delta C Z_0 \sqrt{\epsilon_r}/\eta)^2} \tag{4.40}$$

Substituting $Z_0 = 50\ \Omega$, $\epsilon_r = 2.32$, and the values of ΔC just mentioned, we find that couplers with a coefficient of coupling of less than -12.6 dB (in the former case) and -10.2 dB (in the latter case) can be realized.

Example

Performance specification, $Z_0 = 50\ \Omega$, $c = -15$ dB
Material choice, $b = 3.175$ mm, $t = 350\ \mu$m, $\epsilon_r = 2.32$

From (4.16), $\Delta C = 0.8946$, realizable
From (4.14), $Z_{0e} = 59.845\ \Omega$
From (4.12), $Z_{0o} = 41.774\ \Omega$

These values are inputs to LDIMS and the coupler dimensions are

Line width, $w = 2.277$ mm
Gap, $s = 0.553$ mm

The dimensions of the cross section are shown in Figure 4.6. (For an example and printout of the problem, see also Section B3.2.)

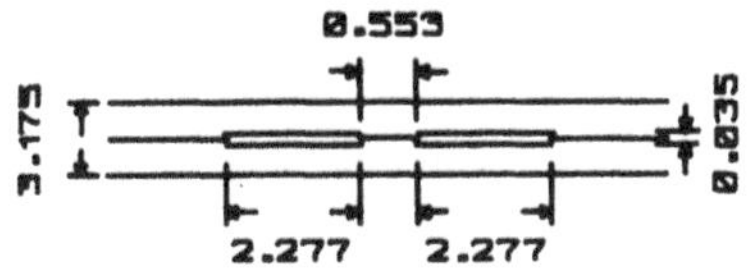

Figure 4.6 Cross section of single-section side-coupled stripline coupler.

4.5.2 Offset Parallel-Coupled Stripline

The offset parallel-coupled stripline can achieve very tight coupling because of the possible proximity of the wide sides of the two strips. The design equations are given in Section B4.1, and these, too, have been programmed into LDIMS. Table B1 shows that it is possible, with conventional, commercially available materials, to achieve coupling values as high as -1.6 dB or mutual capacitances of 7.4. The following design example illustrates the method.

Example

Performance specification, $Z_0 = 50\ \Omega$, $c = -3$ dB
Material choice, $b = 3.30$ mm, $t = 350\ \mu$m, $\epsilon_r = 2.32$
Center strips, $s = 0.125$ mm

From (4.16), $\Delta C = 3.25$, realizable
From (4.14), $Z_{0e} = 120.71\ \Omega$
From (4.12), $Z_{0o} = 20.71\ \Omega$

These values are inputs to LDIMS and the coupler dimensions are

Coupled line width, $w_c = 0.349$ mm
Line width, $w = 1.117$ mm

The cross section is shown in Figure 4.7.

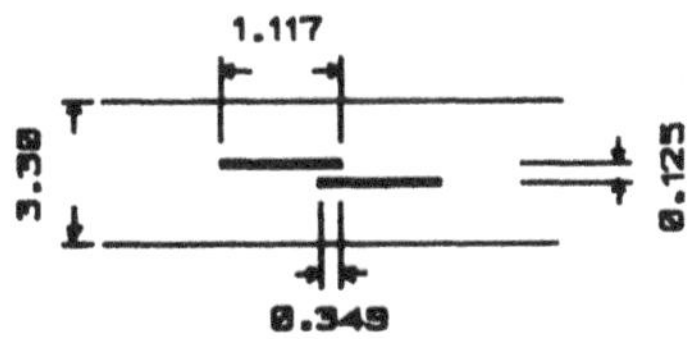

Figure 4.7 Geometry of offset parallel-coupled stripline directional coupler.

4.5.3 Parallel-Coupled Bars

As with the previous type of construction, parallel-coupled bars make possible the realization of very high couplings, due to strong coupling through the thick sides of the bars. Design equations are given in B5.1, and these have also been included in the design program, LDIMS. The program treats the design as if it were a thick stripline. An example of the results using the program follows.

Example

Performance specifications, $Z_0 = 50\ \Omega$, $c = -6$ dB
Material choice, $b = 10$ mm, $t = 4$ mm, $\epsilon_r = 1$

From (4.16), $\Delta C = 4.367$, realizable
From (4.14), $Z_{0e} = 86.74\ \Omega$
From (4.12), $Z_{0o} = 28.28\ \Omega$

These values are inputs to LDIMS and the coupler dimensions are

Line width, $w = 3.208$ mm
Gap, $s = 1.117$ mm

The cross section of the coupler is shown in Figure 4.8.

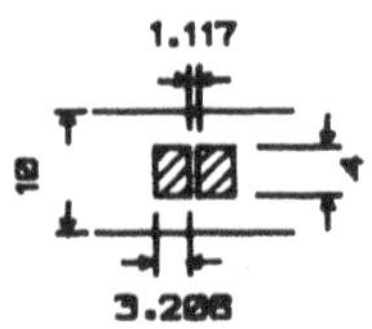

Figure 4.8 **Cross section of a coupler constructed with thick rectangular bars between parallel plates.**

The equations for the realization of parallel coupled bars between parallel plates is in a form that cannot readily be changed for synthesis, so LDIMS in fact keeps on calculating parameters until it finds a set of dimensions that satisfies the given capacitance values. In order to illustrate the mechanisms involved, the example illustrated earlier is now recalculated as an analysis, and the necessary references are given for cases where it is necessary to perform a design without the assistance of a computer.

With $\Delta C = 4.367$, read off from Getsinger ([4], Appendix A, Figure 3), for $t/b = 0.4$:

$s/b = \mathbf{0.11} \qquad s = \mathbf{1.1}$ **mm**

Also from Figure 3, read off ($t/b = 0.4$):

$$C'_{fe}/\epsilon = \mathbf{0.17}$$

and from Figure 5, ($t/b = 0.4$)

$$C'_f/\epsilon = \mathbf{0.92}$$

Using (A.13), calculate:

$$w/b = \tfrac{1}{2}(\tfrac{1}{2}C_{0o}/\epsilon - C'_{fe}/\epsilon - C'_f) = \mathbf{0.326}$$

$$w = \mathbf{3.26}$$

These dimensions are seen to be very close to those calculated by the LDIMS program.

4.5.4 Coupled Microstrip Line

Microstrip line is a nonhomogeneous medium; therefore, the design is slightly more complicated because the parameters change with frequency and the even- and odd-mode velocities differ. Over the years, a large volume of equations for microstrip and coupled microstrip parameters have been published, the latest of which are included in Appendix B and LDIMS.

The following example illustrates the application of LDIMS to the calculation of the dimensions of a −10 dB microstrip coupler. Its dimensions are shown in Figure 4.9.

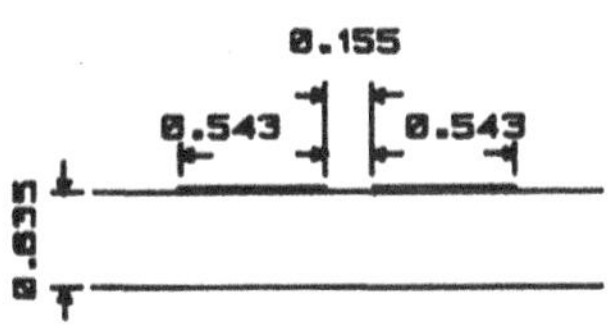

Figure 4.9 Microstrip coupler with −10 dB coupling.

Example

Performance specifications, $Z_0 = 50\ \Omega$, $c = -10$ dB
Material choice, Alumina substrate, $h = 0.635$ mm, $\epsilon_r = 9.6$

From (4.14), $Z_{0e} = 69.37\ \Omega$
From (4.12), $Z_{0o} = 36.34\ \Omega$

These values are inputs to LDIMS and the coupler dimensions are

Spacing, $s = 0.155$ mm
Line width, $w = 0.543$ mm

4.5.5 Lange Coupler

Rizzoli and Lipparini [7] and Paolino [8] present graphical methods for the design of the four coupled lines that are quite simple and rapid to use. However, the design method presented in Osmani [5] can be used to program the design procedure in closed form and is included in LDIMS; the equations are given in Appendix B. The procedure boils down to calculating the even- and odd-mode impedances from (4.37) and (4.38) and using it in the normal calculation for coupled microstrip lines. The following is a typical example of the design of a Lange coupler.

Example

Coupling, $c = -3$ dB
Dielectric constant, $\epsilon_r = 9.6$
Dielectric thickness, $h = .635$ mm
Number of lines, $n = 4$
Characteristic impedance, $Z_0 = 50\ \Omega$

From (4.39), $q = 2.236$
From (4.37), $Z_{0o} = 52.61\ \Omega$
From (4.38), $Z_{0e} = 176.22\ \Omega$

The values for even- and odd-mode impedances were inputs to LDIMS, and the following dimensions were calculated:

Spacing, $s = .0461$ mm
Line width, $w = .0516$ mm

The cross section of the coupler is shown in Figure 4.10.

As a further example, the design of a Lange coupler with six strips on a soft plastic substrate follows. The design procedure is exactly the same as for the case with four strips, except for the value of n.

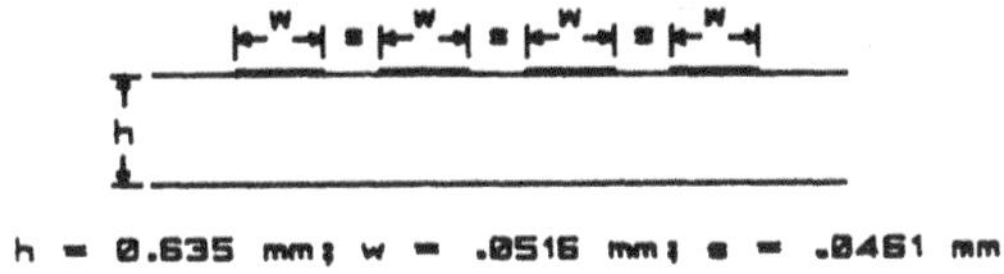

Figure 4.10 **Dimensions of four-strip Lange coupler with −3 dB coupling on alumina substrate.**

Example

Coupling,	$c = -3$ dB
Dielectric constant,	$\epsilon_r = 2.2$
Dielectric thickness,	$h = .508$ mm
Number of lines,	$n = 6$
Characteristic impedance,	$Z_0 = 50\ \Omega$
From (4.39),	$q = 3.6056$
From (4.37),	$Z_{0o} = 82.55\ \Omega$
From (4.38),	$Z_{0e} = 243.1\ \Omega$

The values for even- and odd-mode impedances were inputs to LDIMS, and the following dimensions were calculated:

Spacing,	$s = .0696$ mm
Line width,	$w = .111$ mm

The cross section of the coupler consists of six equal widths of .111 mm and five equal gaps of .0696 mm.

4.6 TANDEM CONNECTION OF DIRECTIONAL COUPLERS

One method of obtaining an increased coupling is by the tandem connection [9] of two directional couplers, as shown in Figure 4.11. For a directional coupler constructed of a pair of coupled lines, this connection entails the disadvantage of two line crossings. But for a Lange coupler, the ports are on the "right" side, so that the two networks can be connected in tandem directly.

The coupling into the coupled port is via both branches $A_1B_1B'_2B'_1$ and $A_1A_2A'_1B'_1$, and thus given by

$$C = j2ct$$

while the path to the through port is given by the difference between $A_1A_2A'_1A'_2$ and $A_1B_1B'_2A'_2$, or

$$T = t^2 - c^2$$

The ratio between the coupled and through ports is then given by

$$\frac{T}{C} = \frac{t^2 - c^2}{j2ct} \tag{4.41}$$

For a pair of -8.34 dB couplers, it can be shown that the power split is equal, or that a -3 dB coupler results.

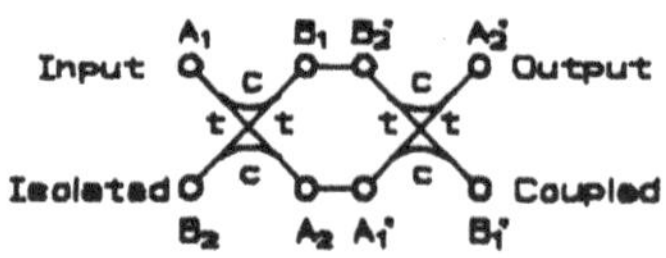

Figure 4.11 **Tandem connection of two directional couplers to give increased coupling.**

REFERENCES

[1] J. A. G. Malherbe, "Application of a Network Model to the Derivation of Coupled Line Properties," submitted for publication.

[2] J. A. G. Malherbe, "Interdigital Directional Couplers with an Odd or Even Number of Lines and Unequal Characteristic Impedances," *Electron. Lett.*, Vol. 12, September 1976, pp. 464–465.

[3] J. Lange, "Interdigitated Stripline Quadrature Hybrid," *IEEE Trans. Microwave Theory Tech.*, Vol. MTT-17, December 1969, pp. 777–779.

[4] R. Waugh and D. LaCombe, "Unfolding the Lange Coupler," *IEEE Trans. Microwave Theory Tech.*, Vol. MTT-20, November 1972, pp. 777–779.

[5] R. M. Osmani, "Synthesis of Lange Couplers," *IEEE Trans. Microwave Theory Tech.*, Vol. MTT-29, February 1981, pp. 168–170.

[6] W. P. Ou, "Design Equations for an Interdigitated Coupler," *IEEE Trans. Microwave Theory Tech.*, Vol. MTT-23, February 1975, pp. 253–255.

[7] V. Rizzoli and A. Lipparini, "The Design of Interdigitated Couplers for MIC Applications," *IEEE Trans. Microwave Theory Tech.*, Vol. MTT-26, January 1978, pp. 7–15.

[8] D. D. Paolino, ''Design More Accurate Interdigitated Couplers,'' *Microwaves,* May 1976, pp. 34–38.

[9] G. B. Monteath, ''Coupled Transmission Lines as Symmetrical Directional Couplers,'' *IEE Proc.*, Part B, May 1955, pp. 383–392.

Chapter 5

Wideband Stepped Coupled-Line Couplers

5.1 INTRODUCTION

In previous chapters we have seen that a considerable increase in bandwidth can be realized by increasing the number of sections making up a directional coupler; in this respect coupled-line couplers are no exception. Some of the principles employed in obtaining wideband prototypes have already been applied, and the problem of the properties of relevant cross sections were studied in detail in the previous chapter, together with the physical realization of such sections.

In this chapter, our first, and very important, step will be to find an equivalent circuit for directional couplers connected in cascade. With such an equivalent circuit, the network properties can be evaluated and a synthesis procedure applied to obtain element values for the resulting network. The method of realization will only be discussed very briefly.

5.2 CASCADE CONNECTION EQUIVALENCE

5.2.1 Equivalent Flow Graph of Two Sections

Figure 5.1(a) shows the signal flow graph for the cascade connection of two single-section couplers. Chapter 4 showed that a coupled-line coupled section will have all its ports matched, providing the coupler condition (4.11) is satisfied. This means that the coupler does not have any inherent mismatch, but no opinion is expressed as to the possible mismatch at any given port due to mismatches at the other ports.

Imposing this latter condition on the flow graph simplifies the graph as shown in Figure 5.1(b); all the reflection coefficients at the ports have disappeared. A surprising fact now emerges. The set of couplers are in

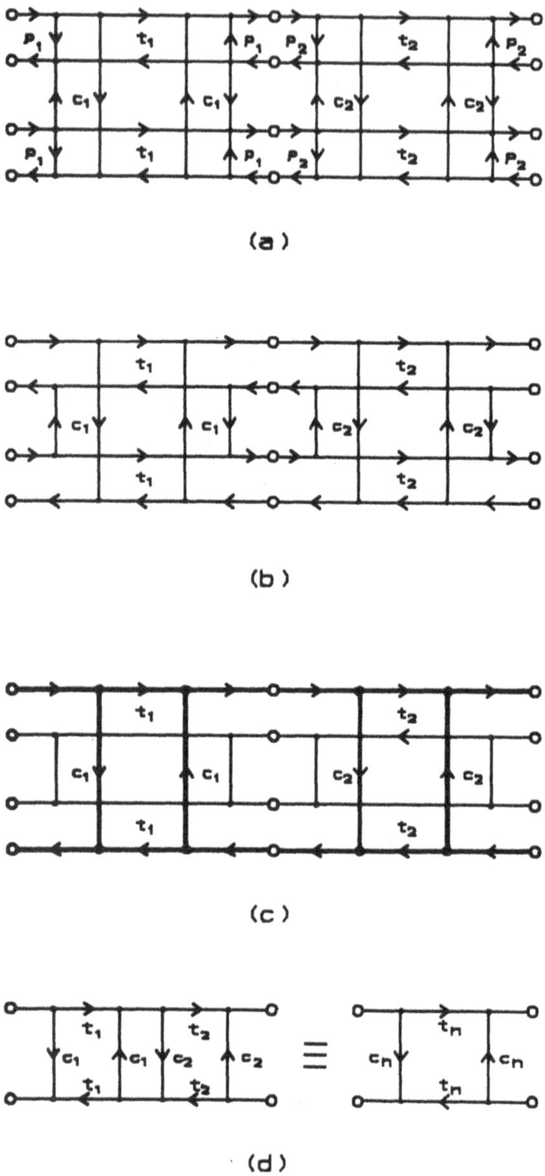

Figure 5.1 Steps in the simplification of two coupled sections.

fact two unconnected two-port networks that will remain unconnected as long as the ports remain matched. In Figure 5.1(c), this is indicated by light and heavy lines.

If, for instance, the heavy lined network is considered separately, as shown in the left-hand side of Figure 5.1(d), it can readily be simplified to the single coupler equivalent of the right-hand side of Figure 5.1(d). Application of flow graph simplification yields

$$t_n = \frac{t_1 t_2}{1 - c_1 c_2} \tag{5.1}$$

$$c_n = c_1 + \frac{t_1^2}{1 - c_1 c_2} \tag{5.2}$$

Providing the coupler condition is still met, this process can be repeated indefinitely; that is, the equivalent circuit of a cascade of coupled-line sections that individually meet the coupler condition will still be a directional coupler with a coupled port in the same position, and 90° out of phase, and with an isolated port. The ports will also be matched at all frequencies. Once again, we express no opinion on the specific properties of the coupled or through power responses. It is the purpose of coupler design to specify the properties of the individual coupled-line sections in such a fashion that those responses will be predetermined.

5.2.2 Unit Element Equivalent Circuit

In the networks of Chapter 3, the even-mode properties of the network were determined through the ratio of the even-mode reflection and transmission coefficients Γ/T. This was necessary because the connected-line couplers and power splitters had at least three parameters that would have to be specified simultaneously, presenting a formidable problem to solve analytically. When the principles were extended in Section 3.4.2 to cover coupled-line power splitters, a new problem arose, as there was suddenly a substantial increase in the number of variables available.

In coupled-line directional couplers, we have shown that (under ideal circumstances):

1. The ports are always matched;
2. There is always total isolation.

The problem of specifying the coupler performance—and the previous section showed that these two principles still apply to multisection couplers—now reduces to specifying only one parameter. The other is automatically specified through conservation of energy.

The coupler condition differentiates coupled-line couplers from the coupled-line power splitters. If the even mode is specified, the whole of the coupler is automatically specified through the coupler condition. Later, we will show that the even mode is appropriate; to this end, consider the input impedance of a unit element of electrical length θ and characteristic impedance Z_u, terminated in a load of Z_0. The input impedance is given by

$$Z_{\text{in}} = Z_u \frac{Z_0 + jZ_u \tan\theta}{Z_u + jZ_0 \tan\theta} \tag{5.3}$$

This corresponds to a reflection coefficient of

$$\Gamma = \frac{\left(Z_u \dfrac{Z_0 + jZ_u \tan\theta}{Z_u + jZ_0 \tan\theta}\right) - Z_0}{\left(Z_u \dfrac{Z_0 + jZ_u \tan\theta}{Z_u + jZ_0 \tan\theta}\right) + Z_0}$$

that after some manipulation can be reduced to

$$\Gamma = \frac{j \left(\dfrac{Z_u^2 - Z_0^2}{Z_u^2 + Z_0^2}\right) \sin\theta}{\left(\dfrac{2Z_u Z_0}{Z_u^2 - Z_0^2}\right) \cos\theta + j \sin\theta} \tag{5.4}$$

If

$$c = \frac{Z_u^2 - Z_0^2}{Z_u^2 + Z_0^2} \tag{5.5}$$

then

$$\Gamma = \frac{jc \sin\theta}{\sqrt{1 - c^2} \cos\theta + j \sin\theta} \tag{5.6}$$

which is the same expression (4.10) derived in Section 4.2.1 for the coupled voltage ratio of a pair of coupled lines. If we apply the principle of conservation of energy, we find the voltage transmission coefficient of the unit element to be

$$T = \frac{\sqrt{1 - c^2}\cos\theta}{\sqrt{1 - c^2}\cos\theta + j\sin\theta} \tag{5.7}$$

which is once again exactly the same as the transmission coefficient for a pair of coupled lines, as given by (4.9).

The implication of (5.6) and (5.7) is vital: The coupling and transmission coefficients of an ideal directional coupler are totally analogous to the voltage reflection and transmission coefficients of a unit element. Furthermore, eqs. (4.13) and (4.12) can be combined to give

$$c = \frac{Z_{0e}^2 - Z_0^2}{Z_{0e}^2 + Z_0^2} \tag{5.8}$$

Comparison with (5.5) shows that the analogy is completed by setting the characteristic impedance of the unit element equal to the even-mode impedance of the coupled-line section. This equivalence is important because there are methods for synthesizing the transfer (and reflection) properties of cascades of unit elements. These results can therefore be applied directly to the synthesis of multisection directional couplers.

5.3 STEPPED ASYMMETRIC COUPLERS

The synthesis of prototypes for realizing symmetric coupled-line couplers follows the same pattern as for multisection power splitters or branch-line couplers. Fortunately, extensive tables of prototypes have been published [1] and the discussion of the method of derivation of the prototypes that follows is given only for the sake of completeness.

5.3.1 Prototype Synthesis Procedure

In Chapter 3, the synthesis problem was approached by first studying the inherent properties of the network and then fitting them to polynomials by equating the coefficients. The same procedure can be employed here, but this method becomes extremely complex for networks above the third order. In this case, the synthesis will start with the desired response function and obtain the network through synthesis after the procedure described by Levy [2].

The energy conservation principle can again be expressed as

$$|\Gamma|^2 + |T|^2 = 1$$

or, rearranging,

$$|\Gamma|^2 = 1 - |T|^2 = 1 - \frac{1}{L} = \frac{L - 1}{L} \tag{5.9}$$

The form that L, the loss function, has to take, is of importance; on the one hand, it must ensure the desired frequency performance, while on the other hand, it must be of a form that can be realized by the network. Because the even-mode network consists entirely of a cascade of unit elements, to satisfy the latter (see 3.2.1), the argument of the variable in the transfer function must be of the form [3]:

$$\cos^2\left[n \cos^{-1}\left(\frac{\sqrt{1 - S_c^2}}{\sqrt{1 - S^2}}\right)\right]$$

Specifically, in terms of the reflection coefficient:

$$|S_{11}|^2 = \frac{|r_{12}|^2}{1 + |r_{12}|^2}$$

so that, for an equiripple directional coupler response:

$$|r_{12}|^2 = \gamma^2 - h^2 \cos^2\left[n \cos^{-1}\left(\frac{\sqrt{1 - S_c^2}}{\sqrt{1 - S^2}}\right)\right] \tag{5.10}$$

where the second term contains a Chebyschev polynomial of the first kind of order n. Figure 5.2 shows the frequency response of the coupled power as well as the reflection coefficient of the equivalent cascade unit element network. The value of the reflection coefficient oscillates between

$$|S_{11}|^2_{\max} = \frac{\gamma^2}{1 + \gamma^2} \tag{5.11}$$

and

$$|S_{11}|^2_{\min} = \frac{\gamma^2 - h^2}{1 + \gamma^2 - h^2} \tag{5.12}$$

At zero frequency, the reflection coefficient is zero, which means that from (5.10)

$$|r_{12}|^2 = 0$$

so that

$$0 = \gamma^2 - h^2 \cos^2 [n \cos^{-1} (\sqrt{1 - S_c^2})]$$

Set

$$S = j \tan\theta, \qquad \sqrt{1 - S^2} = \frac{1}{\cos\theta} \tag{5.13}$$

from which follows, after manipulation,

$$\frac{1}{\cos\theta_0} = \cosh\left(\frac{1}{n}\cosh^{-1}\frac{\gamma}{h}\right) \tag{5.14}$$

The synthesis procedure consists of finding the Chebyschev polynomial of the first kind, and substituting it into (5.10) and (5.9). We then use the reflection coefficient to find the input impedance of the network, from which the actual network elements are obtained. These individual steps will now be studied in more detail.

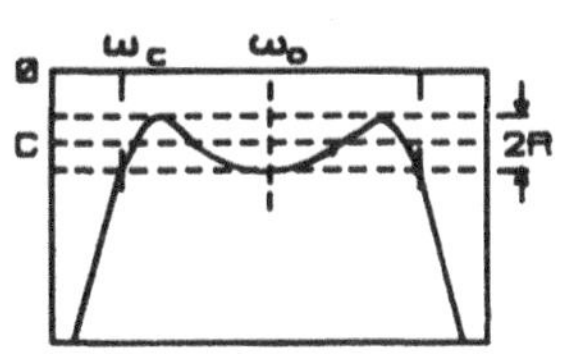

Figure 5.2 Definition of coupler parameters.

The reflection coefficient, Γ, is given by

$$\left|\Gamma\left(\frac{\cos\theta}{\cos\theta_c}\right)\right|^2 = \frac{\gamma^2 - h^2T_n^2\left(\frac{\cos\theta}{\cos\theta_c}\right)}{1 + \gamma^2 - h^2T_n^2\left(\frac{\cos\theta}{\cos\theta_c}\right)} \tag{5.15}$$

The first step in the procedure is to find the roots of the numerator of (5.15). They are given by

$$\cosh\left(n\cosh^{-1}\frac{\cos\theta_r}{\cos\theta_c}\right) = \pm\frac{\gamma}{h}$$

or

$$n\cosh^{-1}\frac{\cos\theta_r}{\cos\theta_c} = \cosh^{-1}\frac{\gamma}{h} + jr\pi, \qquad r = 1, 2, \ldots, n$$

Substituting θ_c from (5.14) gives us

$$\cos\theta_r = \frac{\cosh\,[1/n\cosh^{-1}(\gamma/h + jr\pi/n)]}{\cosh\,[1/n(\cosh^{-1}\gamma/h)]} \tag{5.16}$$

The relationship between $\cos\theta$ and S is known through (5.13), and we can then rewrite (5.16) to give

$$S_r^2 = 1 - \frac{\cosh^2 J/n}{\cosh^2(J/n + jr\pi/n)} \tag{5.17}$$

where

$$J = \cosh^{-1}\gamma/h \tag{5.18}$$

The real and imaginary parts of (5.17) can be readily separated to give us

$$S_r^2 = 1 -$$

$$\left\{ \frac{(\cos^2 r\pi/n)(\cosh^2 J/n) - (\sin^2 r\pi/n)(\sinh^2 J/n) + j\tfrac{1}{2}(\sin 2r\pi/n)(\sinh 2J/n)}{[(\cosh^2 J/n) - (\sin^2 r\pi/n)]^2} \right.$$

$$\left. \cdot \frac{\cosh^2 J/n}{[(\cosh^2 J/n) - (\sin^2 r\pi/n)]^2} \right\} \tag{5.19}$$

The numerator and denominator have the same form. Consequently, if we define,

$$H = \cosh^{-1} \sqrt{1 + \gamma^2}/h \tag{5.20}$$

then the real and the imaginary parts of the roots of the denominator can be written down by inspection as

$$S_r^{2\prime} = 1 -$$

$$\left\{ \frac{(\cos^2 r\pi/n)(\cosh^2 H/n) - (\sin^2 r\pi/n)(\sinh^2 H/n) + j\tfrac{1}{2}(\sin 2r\pi/n)(\sinh 2H/n)}{[(\cosh^2 H/n) - (\sin^2 r\pi/n)]^2} \right.$$

$$\left. \cdot \frac{\cosh^2 J/n}{[(\cosh^2 H/n) - (\sin^2 r\pi/n)]^2} \right\} \tag{5.21}$$

The expression for the power reflection coefficient can finally be written as

$$|\Gamma(S)|^2 = K \frac{(S^2 - S_1^2)(S^2 - S_2^2) \ldots (S^2 - S_n^2)}{(S^2 - S_1^{2\prime})(S^2 - S_2^{2\prime}) \ldots (S^2 - S_n^{2\prime})} \tag{5.22}$$

$$K = \begin{cases} \dfrac{\gamma^2}{1 + \gamma^2} = \left(\dfrac{\cosh J}{\cosh H}\right)^2, & n \text{ odd} \\[2ex] \dfrac{\gamma^2 - h^2}{1 + \gamma^2 - h^2} = \left(\dfrac{\cosh J}{\cosh H}\right)^2, & n \text{ even} \end{cases} \tag{5.23}$$

It is necessary to know the nature of the roots of the numerator (5.22), S_r^2, and the denominator, $S_r^{2\prime}$, to simplify the procedure that follows. The roots are either real, or if complex, occur in conjugate pairs.

For odd orders, there is always one real root for $r = n$, the rest occurring in complex conjugate pairs. The other roots occur as pairs $(r, n - r)$ for $r = 1, 2, \ldots, \frac{1}{2}(n - 1)$. The expression for the power reflection coefficient (5.22) and (5.23) can therefore be rewritten as

$$|\Gamma(S)|^2 = \frac{\gamma}{(1 + \gamma^2)} \frac{(S^2 - S_n^2)}{(S^2 - S_n^{2\prime})} \prod_{r=1}^{\frac{1}{2}(n-1)} \frac{(S^2 - S_r^2)(S^2 - S_r^{2*})}{(S^2 - S_r^{2\prime})(S^2 - S_r^{2\prime *})} \tag{5.24}$$

Any function describing a passive, reciprocal two-port network must have certain properties to prevent nonphysical solutions. The roots of the reflection coefficient of a physical network must have all their poles in the left-hand half of the complex S-plane [3]. In (5.24), the roots in the left-hand half are associated with S_r, and those in the right half of the plane with S_r^*. A further condition, that the resultant network must be a minimum-phase network, demands that the zeros also lie in the left-half plane [1]. Consequently, (5.24) can be rewritten,

$$\Gamma(S) = \frac{\gamma}{\sqrt{(1 + \gamma^2)}} \cdot \frac{(S + S_n)}{(S + S_n')} \prod_{r=1}^{\frac{1}{2}(n-1)} \frac{(S + S_r)(S + S_r^*)}{(S + S_r')(S + S_r'^*)} \tag{5.25}$$

Equation (5.25) is to be used directly in the synthesis procedure and can be simplified by forming the product

$$(S + S_r')(S + S_r'^*) = S^2 + (S_r' - S_r'^*)S + |S_r'|^2 \tag{5.26}$$

The values of the coefficients of S in this expression can be obtained from (5.21). They are

$$|S_r'|^2 = \frac{\sqrt{[\cosh 2(H + J)/n - \cos 2\pi r/n] \cdot [\cosh 2(H - J)/n - \cos 2\pi r/n]}}{(\cosh 2H/n + \cos 2\pi r/n)} \tag{5.27}$$

$$T_r' = S_r' + S_r'^*$$

$$= \sqrt{2\left[1 - \frac{2\cosh^2 J/n(\cosh 2H/n \cos 2r\pi/n + 1)}{(\cosh 2H/n + \cos 2r\pi/n)^2} + |S_r'|^2\right]} \tag{5.28}$$

Due to the symmetry that we noted earlier between the numerator and denominator of (5.15), the following expressions are written by substituting J for H:

$$|S_r|^2 = \frac{\sqrt{[\cosh 2(J+H)/n - \cos 2\pi r/n]\cdot[\cosh 2(J-H)/n - \cos 2\pi r/n]}}{(\cosh 2J/n + \cos 2\pi r/n)} \tag{5.29}$$

$$\begin{aligned} T_r &= S_r + S_r^* \\ &= \sqrt{2\left[1 - \frac{2\cosh^2 H/n\,(\cosh 2J/n \cos 2r\pi/n + 1)}{(\cosh 2J/n + \cos 2r\pi/n)^2} + |S_r|^2\right]} \end{aligned} \tag{5.30}$$

The factors in (5.25) now all have positive real coefficients and will be used in obtaining the coefficients of the cascade matrix for a cascade of unit elements.

For the case of an even number of unit elements, there are two real roots, at $r = n$ and $r = \frac{1}{2}n$, so that the expression for the reflection coefficient is given by

$$\begin{aligned} \Gamma(S) &= \frac{\sqrt{\gamma^2 - h^2}}{\sqrt{1 + \gamma^2 - h^2}} \cdot \frac{(S + S_{n/2})(S + S_n)}{(S + S'_{n/2})(S + S'_n)} \\ &\cdot \prod_{r=1}^{\frac{1}{2}(n-1)} \frac{[S^2 + (S_r + S_r^*)S + |S_r|^2]}{[S^2 + (S'_r + S_r'^*)S + |S'_r|^2]} \end{aligned} \tag{5.31}$$

The input reflection coefficient is now expressed by (5.25) and (5.31) as the quotient of two positive real polynomials of the nth degree and the input inpedance of the network can be obtained from

$$\begin{aligned} Z(S) &= \frac{1 + \Gamma(s)}{1 - \Gamma(S)} \\ &= \frac{P_n(S)}{Q_n(S)} \end{aligned} \tag{5.32}$$

In the next step, the impedance levels of the n lines that make up the stepped impedance prototype network will be determined.

5.3.2 Cascade Network Properties

The cascade matrix of a unit element was defined previously and is given from (1.42) by

$$[A] = \frac{1}{\sqrt{1 - S^2}} \begin{bmatrix} 1 & ZS \\ S/Z & 1 \end{bmatrix}$$

Multiplication of n such sections, for the overall network, gives us

$$[N] = \frac{1}{(1 - S^2)^{n/2}} \prod_{r=1}^{n} \begin{bmatrix} 1 & Z_r S \\ S/Z_r & 1 \end{bmatrix}$$

$$= \frac{1}{(1 - S^2)^{n/2}} \begin{bmatrix} A(S) & B(S) \\ C(S) & D(S) \end{bmatrix} \tag{5.33}$$

From (1.49) and the definition of the reflection coefficient, the input impedance of such a network must be given by

$$Z(S) = \frac{A(S) + B(S)}{C(S) + D(S)} \tag{5.34}$$

where A and D are even and B and C are odd polynomials in S. We can now rewrite eq. (5.32) in the form

$$Z(S) = \frac{E\{P_n(S)\} + O\{P_n(S)\}}{E\{Q_n(S)\} + O\{Q_n(S)\}} \tag{5.35}$$

where $E\{\}$ and $O\{\}$ refer to the even and odd parts of P and Q. Now the two expressions for Z must be the same, so that if these values are substituted in the expression for the network cascade expression (5.33):

$$[N] = \frac{1}{(1 - S^2)^{n/2}} \begin{bmatrix} E\{P_n(S)\} & O\{P_n(S)\} \\ E\{Q_n(S)\} & O\{Q_n(S)\} \end{bmatrix} \tag{5.36}$$

and by (5.33) again,

$$\begin{bmatrix} E\{P_n(S)\} & O\{P_n(S)\} \\ E\{Q_n(S)\} & O\{Q_n(S)\} \end{bmatrix} = \prod_{r=1}^{n} \begin{bmatrix} 1 & Z_r S \\ S/Z_r & 1 \end{bmatrix} \tag{5.37}$$

The characteristic impedances of the cascaded matrices that make up the total network are now obtained by premultiplication by the inverse matrix:

$$\frac{1}{\sqrt{1 - S^2}} \begin{bmatrix} 1 & -Z_1 S \\ -S/Z_1 & 1 \end{bmatrix} \tag{5.38}$$

The value of Z_1 is obtained from the condition that all the elements of the resulting matrix must be divisible by $(1 - S^2)$. This method of extraction is repeated until all the impedances Z_r have been obtained.

If we set the coupling as $C^2_{+,-} = |S_{11}|^2_{\max,\min}$, γ and h are calculated readily from (5.11) and (5.12) as

$$\gamma = \left(\frac{C_+^2}{1 - C_+^2}\right)^{1/2} \tag{5.39}$$

and

$$h = \left(\gamma^2 - \frac{C_-^2}{1 - C_-^2}\right)^{1/2} \tag{5.40}$$

With γ and h and the edge frequencies known, the necessary order of the network can be determined from (5.14) as

$$n = \frac{\cosh^{-1}(\gamma/h)}{\cosh^{-1}(\sec\theta_c)} \tag{5.41}$$

5.3.3 Synthesis Example

Levy [2] illustrates a complete synthesis example for the case of $n = 2$, and we repeated it here as an illustration. The roots given by (5.19) and (5.21) are all real and are given by

$$S_1^2 = \frac{\cosh^2 J/2 + \sinh^2 J/2}{\sinh^2 J/2} = \frac{2\gamma}{\gamma - h} \tag{5.42}$$

$$S_2^2 = 0 \tag{5.43}$$

$$S_1^{2'} = \frac{\sinh^2 H/2 + \cosh^2 J/2}{\sinh^2 H/2} = \frac{\sqrt{1 + \gamma^2} + \gamma}{\sqrt{1 + \gamma^2} - h} \tag{5.44}$$

$$S_2^{2'} = \frac{\cosh^2 H/2 - \cosh^2 J/2}{\cosh^2 H/2} = \frac{\sqrt{1 + \gamma^2} - \gamma}{\sqrt{1 + \gamma^2} + h} \tag{5.45}$$

These values are substituted into (5.31), giving us

$$\Gamma(S) = \frac{\sqrt{\gamma^2 + h^2}}{\sqrt{1 + \gamma^2 - h^2}} \cdot \frac{S\left(S + \sqrt{\dfrac{2\gamma}{\gamma - h}}\right)}{\left(S + \dfrac{\sqrt{1 + \gamma^2} + \gamma}{\sqrt{1 + \gamma^2} - h}\right)\left(S + \dfrac{\sqrt{1 + \gamma^2} - \gamma}{\sqrt{1 + \gamma^2} + h}\right)} \tag{5.46}$$

$$= \frac{\sqrt{\gamma^2 - h^2}\, S^2 + \sqrt{2\gamma(\gamma + h)}\, S}{[\sqrt{1 + \gamma^2 - h^2} S^2 + (\sqrt{(1 + \gamma^2) + (\gamma + h)\sqrt{1 + \gamma^2} + \gamma h} + \sqrt{(1 + \gamma^2) - (\gamma + h)\sqrt{1 + \gamma^2} + \gamma h})S + 1]} \tag{5.47}$$

This expression for $\Gamma(S)$ is substituted in the expression for the input impedance $Z(S)$, in (5.32), and the total cascade matrix of (5.36) can be determined as

$$[N] = \frac{1}{1 - S^2}\begin{bmatrix} aS^2 + 1 & bS \\ cS & dS^2 + 1 \end{bmatrix} \tag{5.48}$$

$$a = \sqrt{1 + \gamma^2 - h^2} + \sqrt{\gamma^2 - h^2} \tag{5.49}$$

$$b = \sqrt{(1 + \gamma^2) + (\gamma + h)\sqrt{1 + \gamma^2} + \gamma h} + \sqrt{(1 + \gamma^2) - (\gamma + h)\sqrt{1 + \gamma^2} + \gamma h} + \sqrt{2\gamma(\gamma + h)} \tag{5.50}$$

$$c = \sqrt{(1 + \gamma^2) + (\gamma + h)\sqrt{1 + \gamma^2} + \gamma h} + \sqrt{(1 + \gamma^2) - (\gamma + h)\sqrt{1 + \gamma^2} + \gamma h} - \sqrt{2\gamma(\gamma + h)} \tag{5.51}$$

$$d = \sqrt{1 + \gamma^2 - h^2} - \sqrt{\gamma^2 - h^2} \tag{5.52}$$

In the next step, we have to obtain the characteristic impedance of the first unit element. We premultiply the cascade matrix of (5.48) by the inverse matrix (5.38) and neglect the $1/(1 - S^2)$ factors for convenience. This gives us

$$\begin{bmatrix} 1 & -Z_1S \\ -S/Z_1 & 1 \end{bmatrix} \begin{bmatrix} aS^2 + 1 & bS \\ cS & dS^2 + 1 \end{bmatrix}$$

$$= \begin{bmatrix} (a - cZ_1)S^2 + 1 & -dZ_1S^3 + (b - Z_1)S \\ -aS^3/Z_1 + (c - a/Z_1)S & (d - b/Z_1)S^2 + 1 \end{bmatrix} \quad (5.53)$$

All the elements of this matrix are divisible by $(1 - S^2)$, giving the two conditions

$$Z_1 = \frac{a + 1}{c} = \frac{b}{d + 1} \quad (5.54)$$

The symmetry condition that was discussed in Chapter 1, (1.36) confirms that the two solutions of (5.54) are in fact the same. Equation (5.53) now reduces to

$$\begin{bmatrix} 1 & \dfrac{dbS}{d + 1} \\ \dfrac{acS}{a + 1} & 1 \end{bmatrix} \equiv \begin{bmatrix} 1 & Z_2S \\ S/Z_2 & 1 \end{bmatrix} \quad (5.55)$$

From (5.55), through inspection follows

$$Z_2 = Z_1/c = dZ_1 \quad (5.56)$$

Equations (5.54) and (5.56) give the element values in terms of the parameters a, b, c, and d, which are in turn all solely a function of γ and h.

Numerical Example

Coupling,	$C = -10$ dB
Ripple,	$R = \pm 0.5$ dB
Bandwidth,	$= 0.5$–2.0 GHz
From (5.39) calculate,	$\gamma = 0.3555$
From (5.40) calculate,	$h = 0.1689$
Center frequency,	$f_0 = 0.5 + \frac{1}{2}(2.0 - 0.5) = 1.25$ GHz
Cutoff angle,	$c = 0.5/1.25 \times 90° = 36°$
Calculate order,	$n = 2.039$ (for the sake of illustration this was taken as 2)

From (5.50) $b = 2.7245$
From (5.52) $d = 0.7350$
From (5.54) $Z_1 = 1.5703$
From (5.56) $Z_2 = 1.1542$

For realization as a coupled line coupler,

$z_{0e1} = Z_1 = 1.5703 \qquad z_{0o1} = 0.6368$

$z_{0e2} = Z_2 = 1.1542 \qquad z_{0o2} = 0.8666$

For a 50 Ω system,

$Z_{0e1} = 78.52\ \Omega \qquad Z_{0o1} = 31.84\ \Omega$

$Z_{0e2} = 57.70\ \Omega \qquad Z_{0o2} = 43.33\ \Omega$

The physical realization of these coupled impedances will be in accordance with the procedures described in Chapter 4.

The even- and odd-mode impedances of this example were fed into SYMFOUR and the coupler responses obtained are shown in Figure 5.3. The coupled power meets the design specification virtually exactly, while the worst theoretical values of return loss and isolation are at -95 dB and -107 dB, respectively.

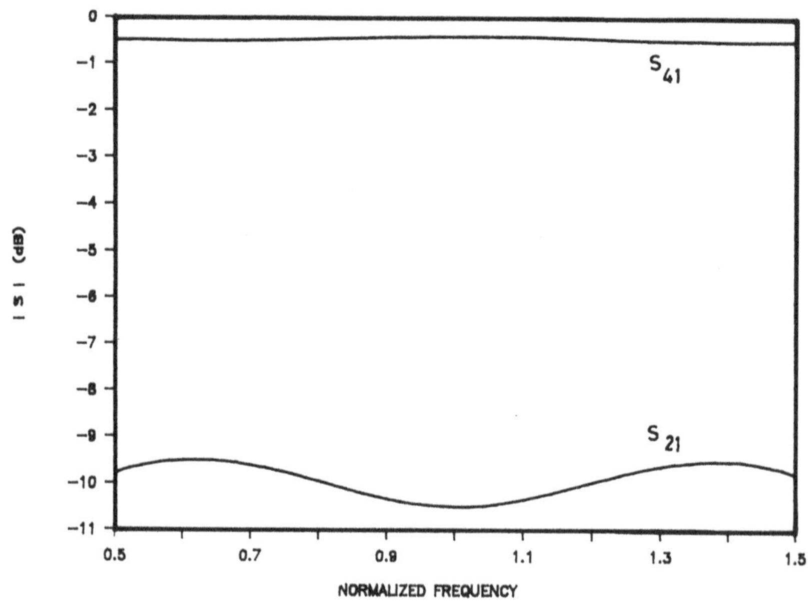

Figure 5.3 **Coupling response *versus* frequency for a 4:1 bandwidth ratio two-section coupled-line coupler.**

For as simple a case as the second order coupler, we have seen that the synthesis procedure is extremely complicated. Levy derives closed-form equations up to and including the fifth order and provides extensive tables [1] of prototype values up to $n = 6$ and for coupling values of -3 dB to -20 dB. These tables are included in Section 5.3.4, together with an illustration of their use.

5.3.4 Tables for Asymmetric Couplers

Tables 5.1–5.5 give the normalized even-mode characteristic impedances of the multisection couplers as presented by Levy [1]. The tables cover designs with $n = 2$ to $n = 6$ elements for a variety of bandwidths and ripple levels. The coupler parameters are defined in Figure 5.4.

Table 5.1 Normalized Even-Mode Impedances ($n = 2$)
(© 1964 IEEE, reprinted by permission, from [1])

BW	*2*	*3*	*4*	*5*	*6*	*7*
$C = -3$ dB						
R (dB)	0.022	0.128	0.292	0.480	0.673	0.858
Z_1	2.9655	3.1292	3.2895	3.4455	3.6007	3.7574
Z_2	1.2308	1.3206	1.4235	1.5320	1.6434	1.7563
$C = -6$ dB						
R (dB)	0.034	0.191	0.439	0.724	1.017	1.303
Z_1	1.9866	2.0596	2.1295	2.1958	2.2603	2.3240
Z_2	1.1481	1.2046	1.2680	1.3335	1.3994	1.4648
$C = -10$ dB						
R (dB)	0.040	0.230	0.528	0.871	1.226	1.575
Z_1	1.5073	1.5412	1.5729	1.6025	1.6307	1.6581
Z_2	1.0882	1.1210	1.1573	1.1940	1.2302	1.2655
$C = -15$ dB						
R (dB)	0.043	0.248	0.568	0.938	1.321	1.698
Z_1	1.2533	1.2688	1.2833	1.2966	1.3091	1.3211
Z_2	1.0480	1.0656	1.0847	1.1038	1.1232	1.1402
$C = -20$ dB						
R (dB)	0.044	0.253	0.580	0.959	1.351	1.737
Z_1	1.1344	1.1423	1.1496	1.1562	1.1624	1.1684
Z_2	1.0266	1.0362	1.0466	1.0569	1.0668	1.0763

Table 5.2 Normalized Even-Mode Impedances ($n = 3$)
(© 1964 IEEE, reprinted by permission from [1])

BW	*4*	*5*	*6*	*7*	*8*	*9*	*10*
$C = -3$ dB							
R (dB)	0.076	0.160	0.266	0.385	0.509	0.635	0.758
Z_1	3.5255	3.6516	3.7672	3.8765	3.9828	4.0881	4.1938
Z_2	1.6386	1.7528	1.8633	1.9072	2.0742	2.1759	2.2759
Z_3	1.1382	1.1925	1.2520	1.3149	1.3801	1.4469	1.5149
$C = -6$ dB							
R (dB)	0.114	0.241	0.400	0.579	0.768	0.959	1.149
Z_1	2.2239	2.2752	2.3215	2.3646	2.4058	2.4462	2.4862
Z_2	1.3875	1.4510	1.5112	1.5683	1.6229	1.6754	1.7263
Z_3	1.0920	1.1276	1.1661	1.2063	1.2475	1.2891	1.3308
$C = -10$ dB							
R (dB)	0.137	0.290	0.482	0.697	0.925	1.156	1.387
Z_1	1.6140	1.6364	1.6564	1.6747	1.6921	1.7089	1.7253
Z_2	1.2215	1.2555	1.2870	1.3166	1.3443	1.3707	1.3959
Z_3	1.0559	1.0772	1.1000	1.1235	1.1473	1.1710	1.1945
$C = -15$ dB							
R (dB)	0.147	0.312	0.518	0.750	0.996	1.246	1.495
Z_1	1.3016	1.3115	1.3203	1.3283	1.3359	1.3431	1.3501
Z_2	1.1174	1.1346	1.1503	1.1649	1.1785	1.1912	1.2033
Z_3	1.0308	1.0424	1.0547	1.0672	1.0798	1.0922	1.1044
$C = -20$ dB							
R (dB)	0.150	0.319	0.530	0.767	1.018	1.274	1.529
Z_1	1.1578	1.1636	1.1680	1.1719	1.1756	1.1792	1.1826
Z_2	1.0642	1.0733	1.0816	1.0892	1.0963	1.1030	1.1092
Z_3	1.0172	1.0236	1.0303	1.0372	1.0440	1.0507	1.0573

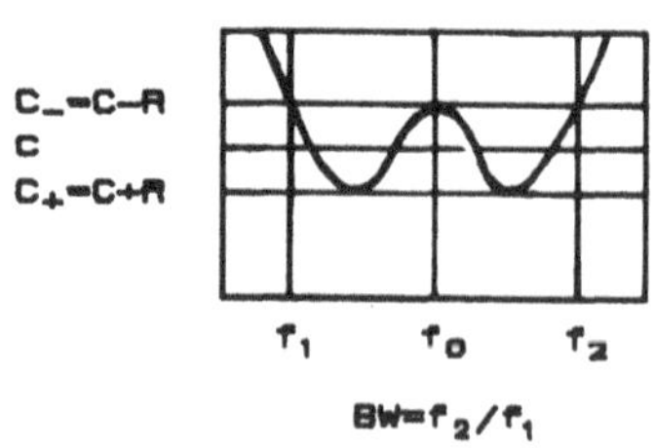

Figure 5.4 Definition of coupler variables. Note that BW is the bandwidth ratio.

Table 5.3 Normalized Even-Mode Impedances ($n = 4$)
(© 1964 IEEE, reprinted by permission, from [1])

BW	4	6	8	10	12	14
$C = -3$ dB						
R (dB)	0.020	0.105	0.250	0.427	0.613	0.796
Z_1	3.7032	3.9210	4.0974	4.2556	4.4088	4.5637
Z_2	1.8149	2.0471	2.2520	2.4379	2.6121	2.7792
Z_3	1.2407	1.3789	1.5184	1.6556	1.7904	1.9232
Z_4	1.0502	1.1108	1.1881	1.2753	1.3683	1.4648
$C = -6$ dB						
R (dB)	0.030	0.158	0.376	0.642	0.925	1.208
Z_1	2.2934	2.3787	2.4459	2.5048	2.5608	2.6166
Z_2	1.4814	1.6028	1.7056	1.7961	1.8786	1.9560
Z_3	1.1561	1.2415	1.3249	1.4047	1.4810	1.5546
Z_4	1.0340	1.0749	1.1262	1.1831	1.2426	1.3031
$C = -10$ dB						
R (dB)	0.036	0.190	0.452	0.773	1.115	1.459
Z_1	1.6438	1.6802	1.7083	1.7326	1.7553	1.7775
Z_2	1.2705	1.3328	1.3840	1.4279	1.4671	1.5032
Z_3	1.0933	1.1424	1.1891	1.2327	1.2735	1.3120
Z_4	1.0210	1.0459	1.0768	1.1105	1.1451	1.1796
$C = -15$ dB						
R (dB)	0.038	0.204	0.487	0.832	1.201	1.572
Z_1	1.3147	1.3306	1.3428	1.3532	1.3629	1.3722
Z_2	1.1419	1.1726	1.1973	1.2181	1.2364	1.2531
Z_3	1.0509	1.0769	1.1012	1.1235	1.1441	1.1632
Z_4	1.0117	1.0254	1.0423	1.0604	1.0788	1.0968
$C = -20$ dB						
R (dB)	0.039	0.209	0.497	0.851	1.229	1.608
Z_1	1.1652	1.1730	1.1790	1.1841	1.1888	1.1933
Z_2	1.0771	1.0932	1.1060	1.1168	1.1261	1.1346
Z_3	1.0282	1.0424	1.0555	1.0675	1.0784	1.0884
Z_4	1.0065	1.0142	1.0235	1.0335	1.0435	1.0533

Table 5.4 Normalized Even-Mode Impedances ($n = 5$)
(© 1964 IEEE, reprinted by permission, from [1])

BW	*6*	*8*	*10*	*12*	*14*	*16*	*18*
$C = -3$ dB							
R (dB)	0.042	0.123	0.239	0.377	0.524	0.673	0.818
Z_1	4.0463	4.2052	4.3384	4.4595	4.5767	4.6945	4.8155
Z_2	2.2021	2.4053	2.5831	2.7439	2.8936	3.0365	3.1751
Z_3	1.4917	1.6410	1.7815	1.9146	2.0424	2.1660	2.2869
Z_4	1.1831	1.2761	1.3726	1.4703	1.5682	1.6659	1.7633
Z_5	1.0510	1.0980	1.1557	1.2206	1.2903	1.3631	1.4381
$C = -6$ dB							
R (dB)	0.062	0.184	0.360	0.567	0.790	1.017	1.241
Z_1	2.4247	2.4839	2.5326	2.5762	2.6178	2.6592	2.7012
Z_2	1.6779	1.7760	1.8592	1.9326	1.9995	2.0622	2.1221
Z_3	1.3064	1.3920	1.4702	1.5423	1.6099	1.6741	1.7356
Z_4	1.1208	1.1800	1.2400	1.2995	1.3580	1.4152	1.4713
Z_5	1.0349	1.0668	1.1055	1.1484	1.1939	1.2406	1.2880
$C = -10$ dB							
R (dB)	0.075	0.222	0.432	0.682	0.952	1.226	1.499
Z_1	1.6992	1.7237	1.7436	1.7612	1.7778	1.7941	1.8104
Z_2	1.3696	1.4173	1.4568	1.4911	1.5218	1.5501	1.5767
Z_3	1.1781	1.2248	1.2664	1.3040	1.3386	1.3709	1.4013
Z_4	1.0730	1.1077	1.1422	1.1759	1.2083	1.2395	1.2696
Z_5	1.0216	1.0411	1.0646	1.0903	1.1171	1.1443	1.1715
$C = -15$ dB							
R (dB)	0.081	0.238	0.465	0.735	1.025	1.321	1.616
Z_1	1.3388	1.3494	1.3579	1.3654	1.3723	1.3791	1.3859
Z_2	1.1902	1.2129	1.2315	1.2473	1.2614	1.2742	1.2862
Z_3	1.0954	1.1193	1.1402	1.1589	1.1758	1.1914	1.2060
Z_4	1.0401	1.0587	1.0770	1.0945	1.1113	1.1272	1.1424
Z_5	1.0120	1.0228	1.0357	1.0496	1.0640	1.0784	1.0927
$C = -20$ dB							
R (dB)	0.082	0.244	0.476	0.751	1.048	1.351	1.653
Z_1	1.1771	1.1822	1.1864	1.1900	1.1934	1.1967	1.1999
Z_2	1.1024	1.1141	1.1236	1.1316	1.1387	1.1452	1.1512
Z_3	1.0524	1.0652	1.0763	1.0861	1.0950	1.1031	1.1106
Z_4	1.0223	1.0325	1.0425	1.0520	1.0610	1.0695	1.0775
Z_5	1.0067	1.0127	1.0199	1.0276	1.0355	1.0433	1.0511

Table 5.5 Normalized Even-Mode Impedances ($n = 6$)
(© 1964 IEEE, reprinted by permission, from [1])

BW	*8*	*10*	*12*	*14*	*16*	*18*	*20*
$C = -3$ dB							
R (dB)	0.060	0.134	0.232	0.344	0.465	0.588	0.711
Z_1	4.2976	4.4178	4.5218	4.6176	4.7105	4.8038	4.8992
Z_2	2.5378	2.7112	2.8640	3.0027	3.1320	3.2551	3.3743
Z_3	1.7504	1.8944	2.0274	2.1520	2.2704	2.3841	2.4944
Z_4	1.3572	1.4618	1.5639	1.6635	1.7606	1.8559	1.9496
Z_5	1.1540	1.2224	1.2946	1.3687	1.4441	1.5200	1.5962
Z_6	1.0523	1.0905	1.1363	1.1875	1.2426	1.3005	1.3603
$C = -6$ dB							
R (dB)	0.090	0.201	0.348	0.518	0.700	0.888	1.077
Z_1	2.5164	2.5596	2.5965	2.6301	2.6623	2.6944	2.7270
Z_2	1.8357	1.9145	1.9821	2.0424	2.0976	2.1494	2.1988
Z_3	1.4506	1.5280	1.5976	1.6613	1.7206	1.7766	1.8301
Z_4	1.2284	1.2913	1.3512	1.4084	1.4632	1.5160	1.5672
Z_5	1.1026	1.1469	1.1928	1.2393	1.2857	1.3318	1.3773
Z_6	1.0360	1.0620	1.0929	1.1271	1.1635	1.2012	1.2397
$C = -10$ dB							
R (dB)	0.109	0.242	0.418	0.623	0.843	1.070	1.299
Z_1	1.7368	1.7543	1.7691	1.7825	1.7952	1.8077	1.8203
Z_2	1.4452	1.4820	1.5130	1.5403	1.5650	1.5878	1.6094
Z_3	1.2554	1.2958	1.3313	1.3633	1.3926	1.4198	1.4455
Z_4	1.1350	1.1704	1.2035	1.2346	1.2639	1.2916	1.3182
Z_5	1.0624	1.0887	1.1155	1.1423	1.1686	1.1944	1.2196
Z_6	1.0223	1.0383	1.0571	1.0778	1.0994	1.1216	1.1440
$C = -15$ dB							
R (dB)	0.117	0.261	0.450	0.670	0.907	1.153	1.400
Z_1	1.3550	1.3624	1.3687	1.3743	1.3796	1.3848	1.3900
Z_2	1.2259	1.2430	1.2573	1.2697	1.2808	1.2910	1.3005
Z_3	1.1346	1.1547	1.1721	1.1876	1.2016	1.2145	1.2266
Z_4	1.0731	1.0915	1.1086	1.1244	1.1392	1.1530	1.1661
Z_5	1.0343	1.0486	1.0629	1.0771	1.0909	1.1043	1.1172
Z_6	1.0124	1.0213	1.0316	1.0429	1.0546	1.0664	1.0783

$C = -20$ dB							
R (dB)	0.119	0.266	0.460	0.685	0.928	1.179	1.431
Z_1	1.1850	1.1886	1.1916	1.1943	1.1969	1.1994	1.2019
Z_2	1.1207	1.1294	1.1366	1.1429	1.1484	1.1535	1.1583
Z_3	1.0733	1.0839	1.0930	1.1011	1.1083	1.1150	1.1211
Z_4	1.0403	1.0503	1.0595	1.0679	1.0758	1.0831	1.0899
Z_5	1.0191	1.0270	1.0348	1.0425	1.0500	1.0572	1.0642
Z_6	1.0070	1.0119	1.0176	1.0239	1.0303	1.0368	1.0433

Design Example

The following design example illustrates the use of the tables as well as the theoretical performance achieved.

Coupling, $C = -6$ dB
Bandwidth, $BW = 10:1$
Ripple level, $R \leq \pm.75$ dB

A coupler that would satisfy these specifications is given in Table 5.4, $n = 4$, $C = 6$ dB, $R = 0.642$ dB.

$Z_1 = \mathbf{2.5048}$ $Z_2 = \mathbf{1.7961}$

$Z_3 = \mathbf{1.4047}$ $Z_4 = \mathbf{1.1831}$

The following values are calculated for these sections, for a 50 Ω system. The even- and odd-mode impedances for each section are given, together with the coupling coefficient and the mutual capacitance using a dielectric with $\epsilon_r = 2.32$:

n	Z_{0e}	Z_{0o}	c	ΔC
1	125.24	9.96	0.7250	5.2115
2	89.81	27.84	0.5267	3.0675
3	70.24	35.59	0.3237	1.7148
4	59.16	42.26	0.1666	0.8363

The values of ΔC show that offset parallel-coupled stripline would have to be used to realize the first two sections; the whole network is consequently realized in this medium, with

ϵ_r = 2.32 b = 3.3 mm
center strip ratio = 0.03788

The even- and odd-mode impedances in this table were inputs to LDIMS and the following dimensions calculated:

n	w_c	w
1	0.3845	1.0717
2	0.1021	1.5850
3	−0.1204	2.1137
4	−0.5732	2.4700

These even- and odd-mode impedances were used in SYMFOUR to calculate the responses shown in Figure 5.5. The coupling is as specified; the return loss and isolation have been shown on the same graph and, in both cases, are greater than 80 dB.

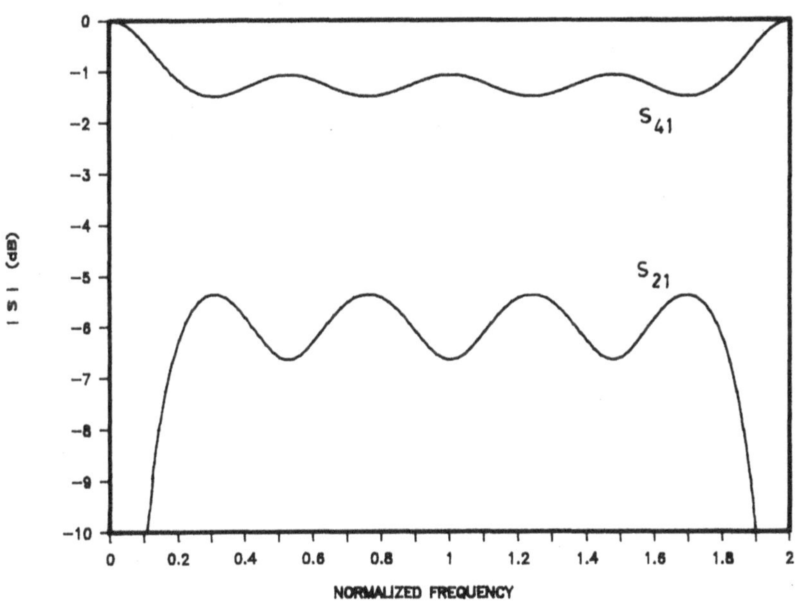

Figure 5.5 Theoretical responses of *S*-parameters for the four-section design example.

5.4 STEPPED SYMMETRICAL COUPLERS

In this section we follow the same procedure for determining a set of prototypes for the design of symmetrical couplers that we used for designing asymmetical couplers. As earlier, the section will conclude with tables of element values for the design of couplers as well as a design example. The synthesis and design procedure is based on yet another classic paper, by Cristal and Young [4]; in principle it is the same as all the previous ones, except that in its specific execution it follows a pattern that is most convenient for the application.

The synthesis approach utilized in the previous section can be described as a direct approach. It used the synthesis procedure for conventional minimum phase filter networks, with conventional Chebyschev polynomials to describe the desired equiripple function. As the minimum of specifications had been set, the resultant network was asymmetric. In this case, the insertion loss function will be specified in such a way that the resultant physical network will be symmetrical. This condition is achieved [4] by using a special equiripple function. In order not to interrupt the flow of the argument, this equiripple function will first be discussed briefly.

5.4.1 Special Equiripple Polynomials

In order to obtain a symmetrical network, we have seen that its transfer function must be of the form [4]

$$|r_{12}|^2 = [P_n(\sin\theta)]^2 \tag{5.57}$$

where P_n is an odd polynomial in $\sin\theta$ of degree n. This polynomial is not the conventional Chebyschev polynomial used in the previous section. Furthermore, the variable $\sin\theta$ is a low-pass (bandstop) variable [3] and, consequently, P_n will have to be a high-pass function as opposed to the conventional Chebyschev, which is a low-pass function.

As the complete procedure for obtaining the polynomial is described in [4] and it is unlikely that most readers will need to construct such polynomials in view of the extensive tables, we will discuss only some of its properties here. As an example, Figure 5.6 shows the desired ripple properties of the polynomial $P_9(x)$ across its variable range of $0 \leq x \leq 1$. The function will reach a maximum value of P^+ twice and have its last equiripple value of P^+ at $x = 1$. It will also have two minima of P^-. For this specific case, and for a ripple of $R = 0.1$, the polynomial is given by [4] as

$$P_9(x) = 7.238206x - 47.4794411x^3 + 138.138390x^5 - 164.008299x^7 + 68.311141x^9 \tag{5.58}$$

An important property of the polynomials is that they all have a zero at $x = 0$, because they are odd. This ensures the desired high-pass property and makes the network physically realizable. Beyond $x = 1$, the function keeps on increasing, but this region lies outside of the range of $\sin\theta$.

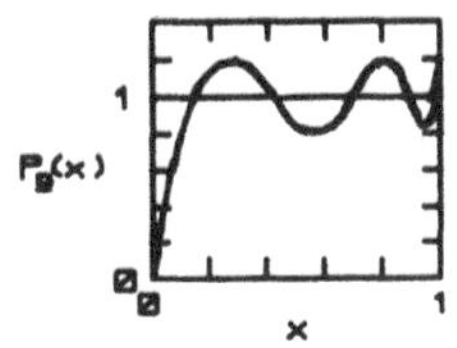

Figure 5.6 Defining the properties of the equiripple polynomial, in this case $P_9(x)$.

5.4.2 Prototype Synthesis Procedure

Once the polynomial $P_n(x)$ is known, we can perform the synthesis of the impedance function. The reflection coefficient, from (5.9), is given as

$$|S_{11}|^2 = \frac{|r_{12}|^2}{1 + |r_{12}|^2} = \frac{P_n^2(x)}{1 + P_n^2(x)} \tag{5.59}$$

The substitution

$$x \leftrightarrow s/j \tag{5.60}$$

is now made, where s is an intermediate variable and not yet the Richards' variable, $S = j\tan\theta$. The right-hand side of (5.59) can now be factored into the form

$$|\Gamma|^2 = \Gamma(s/j)\Gamma(-s/j) \propto \frac{\prod_{i=1}^{n} (s - z_i)(-s - z_i)}{\prod_{i=1}^{n} (s - p_i)(-s - p_i)} \tag{5.61}$$

where z_i are the zeros of the numerator and p_i are the zeros of the denominator. The reason for the proportionality rather than equality in the right-hand side of (5.61) is that there are still constants missing in (5.61).

The zeros are obtained by solving

$$P_n(s/j) = 0 \tag{5.62}$$

and

$$P_n(s/j) = \pm j \tag{5.63}$$

Note that, as P_n is odd, only one of the equations in (5.63) has to be solved, because the solutions with the positive sign on the right-hand side are the negative of the values with the negative sign.

The substitution

$$s \leftrightarrow S = j \tan\theta \tag{5.64}$$

is now made, and the zeros obtained earlier are mapped into new zeros

$$Z_i = \frac{z_i}{\sqrt{1 + (z_i)^2}} \tag{5.65}$$

$$P_i = \frac{p_i}{\sqrt{1 + (p_i)^2}} \tag{5.66}$$

The choice of the poles and zeros associated with the function $\Gamma(S)$ is in accordance with normal procedure. To ensure an analytic solution, the zeros of the denominator of Γ must lie in the left-half plane. In order to ensure that the network will be symmetrical, the zeros of the numerator must be symmetrical with respect to the imaginary axis. The complex reflection coefficient is now constructed as

$$\Gamma(S) = \frac{\gamma_N(S)}{\gamma_D(S)} = \frac{a \prod_{i=1}^{n} (S - Z_i)}{b \prod_{i=1}^{n} (S - P_i)} \tag{5.67}$$

In effect, we have now made the single transform:

$$x = \frac{(S/j)}{\sqrt{1 + (S/j)^2}} \tag{5.68}$$

In order to eliminate the constants a and b, we determine the values of the numerator and denominator at $S = \sqrt{2}$. Substituting into (5.68) gives $x = \sqrt{2}$, and comparing the denominators in (5.61) and (5.59) gives

$$P_n^2(\sqrt{2}) = a^2(-1)^{-n} \prod_{i=1}^{n} (\sqrt{2} - Z_i)(-\sqrt{2} - Z_i) \tag{5.69}$$

and after solving for a

$$a = \left[\frac{(-1)^n P_n^2(\sqrt{2})}{\prod_{i=1}^{n} (\sqrt{2} - Z_i)(-\sqrt{2} - Z_i)} \right]^{1/2} \tag{5.70}$$

Cristal and Young [4] argue that the positive root must be taken in (5.70), so that the $\gamma_N(S) \to +0$ as $S \to 0$. This choice is justified by noting that the normalized even-mode impedance is always greater than one, which requires the reflection coefficient to be positive as $S \to 0$. By a similar argument, the constant in the denominator is obtained as

$$b = \left[\frac{(-1)^n [1 + P_n^2(\sqrt{2})]}{\prod_{i=1}^{n} (\sqrt{2} - P_i)(-\sqrt{2} - P_i)} \right]^{1/2} \tag{5.71}$$

We now know the complex reflection coefficient $\Gamma(S)$ from eqs. (5.67), (5.70), and (5.71) and can determine the input impedance from

$$Z(S) = \frac{1 + \Gamma(S)}{1 - \Gamma(S)} \tag{5.72}$$

This brings the synthesis procedure to the same point as (5.32) in Section 5.3.1.

5.4.3 Network Synthesis

We have now determined the *ABCD* matrix of the network and can use the same procedure as before to determine the network properties. Cristal and Young [4] explain very elegantly why the polynomial $P_n(x)$ must be odd to ensure a symmetrical network. Consider the resultant cascade matrix. We see that the element A is the even part of the sum of the numerator and denominator of P_n:

$$A = E\{\gamma_D + \gamma_N\}$$

while the element D is the even part of the differences:

$$D = E\{\gamma_D - \gamma_N\}$$

For symmetry, by (1.36), $A = D$. The only way in which this can be achieved is by making γ_N an odd polynomial by choosing a zero at the origin, and arranging the other zeros symmetrically on either side of the imaginary axis. For the synthesis of the final even-mode network, the previous procedure is used.

5.4.4 Tables for Symmetrical Couplers

Rather than performing the procedure just described, we design a directional coupler by using the extensive tables published in [4]. As the procedure is identical to that in Section 5.3, only the tables are given. Note that only one half of the prototype is listed; the other half is obtained by making use of symmetry. We use the same nomenclature as in 5.3.4.

Table 5.6 Normalized Even-Mode Impedances ($n = 3$; $Z_3 = Z_1$) (© 1965 IEEE, reprinted by permission from [4]).

R (dB)	*0.1*	*0.2*	*0.4*	*0.6*	*0.8*	*1.0*
$C = -3$ dB						
BW	3.03063	3.83085	5.17521	6.45616	7.77966	9.20361
Z_1	1.17135	1.20776	1.27036	1.32964	1.38970	1.45274
Z_2	3.25984	3.41242	3.66560	3.90585	4.15648	4.43120
$C = -6$ dB						
BW	2.70356	3.31984	4.29931	5.17291	6.01830	6.86621
Z_1	1.10298	1.12090	1.15038	1.17680	1.20208	1.22698
Z_2	2.09445	2.14693	2.22865	2.29968	2.36724	2.43431
$C = -8.34$ dB						
BW	2.61290	3.18211	4.07347	4.85550	5.60103	6.33787
Z_1	1.07434	1.08644	1.10606	1.12339	1.13973	1.15560
Z_2	1.71858	1.74864	1.79461	1.83365	1.86993	1.90510
$C = -10$ dB						
BW		3.12968	3.98852	4.73738	5.44739	6.14545
Z_1		1.06945	1.08475	1.09817	1.11075	1.12290
Z_2		1.57423	1.60708	1.63470	1.66014	1.68458
$C = -20$ dB						
BW		3.03958	3.84396	4.53804	5.19011	5.82570
Z_1		1.02070	1.02497	1.02866	1.03208	1.03534
Z_2		1.14915	1.15617	1.16197	1.16720	1.17213

Table 5.7 Normalized Even-Mode Impedances ($n = 5$; $Z_5 = Z_1$; $Z_4 = Z_2$) (© 1965 IEEE, reprinted by permission, from [4])

R (dB)	*0.1*	*0.2*	*0.4*	*0.6*	*0.8*	*1.0*
$C = -3$ dB						
BW	4.93114	6.29714	8.55845	10.69292	12.88720	15.24047
Z_1	1.07851	1.10921	1.16266	1.21370	1.26555	1.31998
Z_2	1.37268	1.44029	1.54541	1.63864	1.73013	1.82466
Z_3	3.97615	4.21023	4.57491	4.90924	5.25363	5.62978
$C = -6$ dB						
BW	4.34522	5.45738	7.08866	8.55462	9.96442	11.37370
Z_1	1.04501	1.06052	1.08633	1.10969	1.13217	1.15438
Z_2	1.21972	1.25302	1.30203	1.34262	1.37978	1.41542
Z_3	2.38181	2.46010	2.57332	2.66727	2.75470	2.84048
$C = -8.34$ dB						
BW	4.20727	5.19150	6.70767	8.02323	9.26959	10.49635
Z_1	1.03211	1.04271	1.06012	1.07565	1.09039	1.10476
Z_2	1.15690	1.17918	1.21142	1.23760	1.26114	1.28331
Z_3	1.89019	1.93414	1.99635	2.04658	2.09210	2.13562
$C = -10$ dB						
BW		5.10148	6.56407	7.82513	9.01322	10.17639
Z_1		1.03418	1.04784	1.05996	1.07140	1.08249
Z_2		1.14316	1.16808	1.18815	1.20606	1.22280
Z_3		1.70922	1.75305	1.78805	1.81943	1.84912
$C = -20$ dB						
BW		4.94656	6.31936	7.49038	8.58338	9.64410
Z_1		1.01016	1.01406	1.01747	1.02066	1.02371
Z_2		1.04183	1.04855	1.05386	1.05851	1.06280
Z_3		1.17873	1.18767	1.19463	1.20073	1.20638

Table 5.8 Normalized Even-Mode Impedances ($n = 7$; $Z_{8-i} = Z_i$)
(© 1965 IEEE, reprinted by permission, from [4])

R (dB)	*0.1*	*0.2*	*0.4*	*0.6*	*0.8*	*1.0*
$C = -3$ dB						
BW	6.9531	8.8860	12.0666	15.0578	18.1270	21.4147
Z_1	1.05240	1.07950	1.12798	1.17505	1.22323	1.27399
Z_2	1.18406	1.23581	1.31754	1.39069	1.46258	1.53668
Z_3	1.56753	1.65795	1.79367	1.91172	2.02682	2.14566
Z_4	4.61180	4.90662	5.35611	5.76434	6.18437	6.64407
$C = -6$ dB						
BW	6.1494	7.6583	10.0026	12.0626	14.0396	16.0119
Z_1	1.02686	1.04246	1.06580	1.08735	1.10831	1.12915
Z_2	1.10756	1.13419	1.17408	1.20755	1.23839	1.26805
Z_3	1.32930	1.37278	1.43418	1.48367	1.52841	1.57104
Z_4	2.62516	2.72038	2.85438	2.96391	3.06523	3.16446
$C = -8.34$ dB						
BW	5.9117	7.3140	9.4572	11.3077	13.0563	14.7747
Z_1	1.02033	1.02963	1.04538	1.05972	1.07350	1.08702
Z_2	1.07694	1.09518	1.12204	1.14417	1.16423	1.18319
Z_3	1.23301	1.26167	1.30136	1.33267	1.36040	1.38629
Z_4	2.03194	2.08436	2.15641	2.21361	2.26506	2.31408
$C = -10$ dB						
BW		7.1965	9.2638	11.0383	12.7059	14.3359
Z_1		1.02360	1.03597	1.04718	1.05786	1.06834
Z_2		1.07622	1.09725	1.11444	1.12991	1.14444
Z_3		1.20802	1.23839	1.26213	1.28298	1.30229
Z_4		1.81699	1.86715	1.90649	1.94149	1.97446
$C = -20$ dB						
BW		6.9766	8.9188	10.5678	12.1029	13.5903
Z_1		1.00697	1.01052	1.01369	1.01669	1.01958
Z_2		1.02256	1.02846	1.03320	1.03740	1.04129
Z_3		1.05976	1.06767	1.07372	1.07894	1.08368
Z_4		1.20128	1.21112	1.21863	1.22515	1.23116

Table 5.9 Normalized Even-Mode Impedances ($n = 9$; $Z_{10-i} = Z_i$)
(© 1965 IEEE, reprinted by permission, from [4])

R (dB)	*0.1*	*0.2*	*0.4*	*0.6*	*0.8*	*1.0*
$C = -3$ dB						
BW	9.030	11.528	15.627	19.475	23.421	27.644
Z_1	1.04112	1.06598	1.11149	1.15624	1.20234	1.25107
Z_2	1.12024	1.16366	1.23397	1.29789	1.36117	1.42660
Z_3	1.29488	1.36260	1.46548	1.55536	1.64277	1.73250
Z_4	1.74863	1.85696	2.01711	2.15551	2.29038	2.42995
Z_5	5.18240	5.52654	6.04655	6.51769	7.00316	7.53602
$C = -6$ dB						
BW	7.989	9.943	12.969	15.622	18.166	20.702
Z_1	1.02201	1.03437	1.05615	1.07658	1.09661	1.11663
Z_2	1.06888	1.09137	1.12599	1.15561	1.18320	1.20989
Z_3	1.17282	1.20736	1.25686	1.29716	1.33370	1.36852
Z_4	1.42807	1.47877	1.54902	1.60504	1.65546	1.70342
Z_5	2.83542	2.94305	3.09269	3.21427	3.32658	3.43663
$C = -8.34$ dB						
BW	7.681	9.498	12.265	14.650	16.901	19.111
Z_1	1.01536	1.02379	1.03846	1.05206	1.06523	1.07823
Z_2	1.04904	1.06452	1.08798	1.10771	1.12579	1.14302
Z_3	1.12341	1.14687	1.17989	1.20622	1.22966	1.25158
Z_4	1.30048	1.33341	1.37809	1.41290	1.44356	1.47211
Z_5	2.15200	2.21025	2.28925	2.35152	2.40740	2.46063
$C = -10$ dB						
BW		9.345	12.016	14.303	16.450	18.547
Z_1		1.01889	1.03041	1.04103	1.05127	1.06133
Z_2		1.05161	1.07004	1.08543	1.09945	1.11271
Z_3		1.11743	1.14313	1.16344	1.18139	1.19805
Z_4		1.26387	1.29777	1.32390	1.34672	1.36779
Z_5		1.90628	1.96074	2.00313	2.04073	2.07614
$C = -20$ dB						
BW		9.061	11.571	13.697	15.674	17.588
Z_1		1.00555	1.00886	1.01187	1.01474	1.01753
Z_2		1.01529	1.02054	1.02485	1.02871	1.03232
Z_3		1.03447	1.04153	1.04700	1.05175	1.05608
Z_4		1.07471	1.08328	1.08974	1.09527	1.10028
Z_5		1.21931	1.22965	1.23748	1.24426	1.25049

REFERENCES

[1] R. Levy, "Tables for Asymmetric Multi-Element Coupled-Transmission-Line Directional Couplers," *IEEE Trans. Microwave Theory Tech.*, Vol. MTT-12, May 1964, pp. 275–279.
[2] R. Levy, "General Synthesis of Asymmetric Multi-Element Coupled-Transmission-Line Directional Couplers," *IEEE Trans. Microwave Theory Tech.*, Vol. MTT-11, July 1963, pp. 226–237.
[3] J. A. G. Malherbe, Microwave Transmission Line Filters, Artech House, Dedham, MA, 1979.
[4] E. G. Cristal, and L. Young, "Theory and Tables of Optimum Symmetrical TEM-Mode Coupled-Transmission-Line Directional Couplers," *IEEE Trans. Microwave Theory Tech.*, Vol. MTT-13, September 1965, pp. 544–558.

Chapter 6

Ultrawideband Couplers with Continuously Varying Cross Section

6.1 INTRODUCTION

In this chapter, the coupled-line couplers have coupled transmission lines of continuously varying cross section. For coupled-line couplers, we have used an equivalence principle that equates the reflection coefficient of a cascade of unit elements to the coupling of corresponding stepped-coupled line sections. Consequently, we could solve the much simpler problem of cascaded unit elements and then apply the results to coupler design.

Sharpe [1] shows that a similar equivalence principle holds for nonuniform coupled lines. This means that we can design nonuniform coupled-line directional couplers without having to assume that the coupling is small or that the lines are only slightly nonuniform. The interested reader is referred directly to Sharpe's paper [1]; it will not be discussed here. In an earlier work [2], Sharpe discusses the method used by Orlov to obtain an expression for the reflection coefficient of a continuously varying transmission line. The method involves the approximation of an arbitrary nonuniform line by a succession of equal-length steps. Starting at the load, he obtains in the limit the input reflection coefficient at one step as the output reflection coefficient of the next step. Allowing the length of the steps to approach zero, the input reflection coefficient is obtained in the limit.

The reflection coefficient of a transmission line of length l and characteristic impedance in the point x of $Z_0(x)$ can be expressed as

$$\rho e^{-j\beta l} = \int_{-\frac{1}{2}l}^{\frac{1}{2}l} e^{-j2\beta x} p(x)\, dx \tag{6.1}$$

where

$$p(x) = \tfrac{1}{2} \frac{d}{dx} \ln Z_0(x) \tag{6.2}$$

By the equivalence principle, the equations can be rewritten for coupled lines as

$$ce^{-j\beta l} = \int_{-\frac{1}{2}l}^{\frac{1}{2}l} e^{-j2\beta x} \, p(x) \, dx \tag{6.3}$$

where

$$p(x) = \tfrac{1}{2} \frac{d}{dx} \ln Z_{0e}(x) \tag{6.4}$$

Through the application of the Fourier transform, (6.1) and (6.2) can be rewritten as

$$F(x) = \frac{1}{\pi} \int_{-\infty}^{\infty} \left| \rho e^{j\beta l} \right| e^{-j2\beta x} \, d\beta \tag{6.5}$$

These equations form the basis of a variety of solutions to the problem of nonuniform coupled-line couplers. (As all the couplers under consideration in this chapter make use of nonuniform lines, we will no longer refer to them as such; rather, when we use uniform coupled lines we will state so.) The various methods differ enough to make a unified treatment difficult, and so the analysis and design for the different approaches will split at this early state. To summarize them in advance might be confusing, so we will only develop a perspective as the theory develops.

6.2 TAPERED-LINE MAGIC-T

The tapered-line magic-T [3] is based on a tapered asymmetrical transformer. Physically, it consists of the two coupled tapered lines, cascaded with two uncoupled lines of the same length, which are necessary to correct the phase of the final coupler. Contrary to the initial (and obvious) supposition the tapered-line magic-T is not the nonuniform line equivalent of the stepped asymmetric coupler that we treated in Section 5.3. Rather, it is based on the principle that there is a perfect transmission-line transformer that transforms the port characteristic impedance (taken as unity throughout) to an even-mode impedance of $1/k$; at the end of this transformer is a single reflection, which will of course correspond to the coupling, when the wave meets the unit characteristic impedance at the opposite port.

The well-known coupler condition is defined as

$$z_{0e}(x)z_{0o}(x) = 1 \tag{6.6}$$

which means that for the odd mode, the impedance transformation will be to an impedance level of k, of course.

6.2.1 Equivalent Circuit

The structure to be analyzed is shown in Figure 6.1(a) and the equivalent even- and odd-mode networks in 6.1(b) and 6.1(c) respectively. The coupled-line section has been replaced by a transmission line of unit impedance, together with a transformer of turns ratio $\sqrt{k}$ for the even mode and a turns ratio of $1/\sqrt{k}$ for the odd mode. Consider next the equivalent circuit of a network having a transformer with a turns ratio of n, cascaded in between two transmission lines of unit impedance. The cascade matrix for such a network is given by

$$[A] = \frac{1}{\sqrt{1 - S^2}} \begin{bmatrix} 1 & SZ_0 \\ S/Z_0 & 1 \end{bmatrix} \begin{bmatrix} n & 0 \\ 0 & 1/n \end{bmatrix} \frac{1}{\sqrt{1 - S^2}} \begin{bmatrix} 1 & SZ_0 \\ S/Z_0 & 1 \end{bmatrix} \tag{6.7}$$

and if we multiply this out and set $Z_0 = 1$:

$$[A] = \frac{1}{1 - S^2} \begin{bmatrix} S^2 + n^2 & S(1 + n^2) \\ S(1 + n^2) & n^2S^2 + 1 \end{bmatrix} \tag{6.8}$$

In terms of this cascade matrix, the reflection coefficient, from (1.49), is given by

$$\Gamma = \frac{A + B - C - D}{A + B + C + D} \tag{6.9}$$

Setting $n = \sqrt{k}$ for the even mode, and $n = 1/\sqrt{k}$ for the odd mode, the odd and even reflection coefficients are readily obtained from (6.3), and then substituted into (1.45) and (1.46) to obtain

$$S_{11} = \tfrac{1}{2}(\Gamma_e + \Gamma_o) = 0 \tag{6.10}$$

$$
\begin{aligned}
S_{12} &= \tfrac{1}{2}(\Gamma_e - \Gamma_o) \\
&= \frac{1}{2}\left(-\frac{1-k}{1+k} + \frac{1-1/k}{1+1/k}\right)\frac{(1-S)}{(1+S)} \\
&= -\frac{1-k}{1+k}\cdot\frac{(1-S)}{(1+S)} \\
&= -\frac{1-k}{1+k}e^{-j2\theta} \qquad (6.11)
\end{aligned}
$$

Similarly, the transmission coefficients can be obtained from (1.47), (1.48), and (1.50):

$$S_{13} = 0 \qquad (6.12)$$

$$S_{14} = -\frac{2\sqrt{k}}{1+k}e^{-j2\theta} \qquad (6.13)$$

$$S_{44} = 0 \qquad (6.14)$$

$$S_{43} = \frac{1-k}{1+k}e^{-j2\theta} \qquad (6.15)$$

Note that there is a phase reversal between S_{14} and S_{43}. The maximum value of the coupling is readily determined from (6.11) as

$$k_{\max} = \frac{1-|S_{12}|}{1+|S_{12}|} \qquad (6.16)$$

The ratio

$$\rho_0 = \frac{1-k}{1+k} = c_0 \qquad (6.17)$$

is the nominal coupling of the coupler and corresponds to the nominal reflection coefficient of the transformer when connected to a unit load. The variation in reflection coefficient under these conditions is then bounded as

$$\frac{\rho_0 - \rho_m}{1 - \rho_0\rho_m} \le |S_{12}| \le \frac{\rho_0 + \rho_m}{1 + \rho_0\rho_m} \qquad (6.18)$$

where ρ_m is the maximum ripple of the reflection coefficient of the transformer. What remains now is to find a taper that is an ideal transformer between unit impedance and $1/k$ for the even mode of the coupled-line formulation.

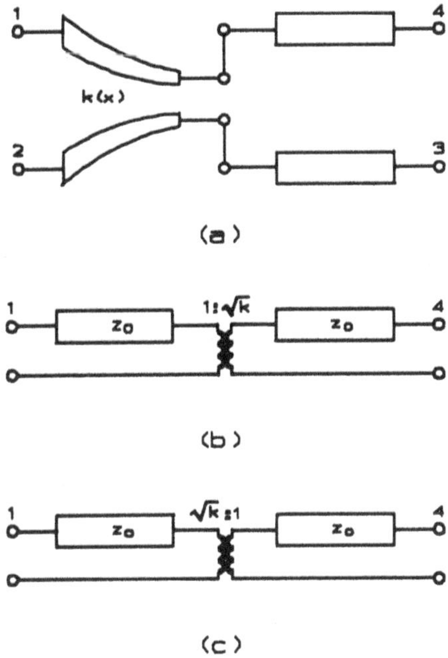

Figure 6.1 Development of the equivalent circuit of a tapered-line magic-T shown in (a). The even-mode network is shown in (b), and the odd-mode network in (c).

6.2.2 Tapered-Line Transformer

The paper by DuHamel and Armstrong [3] originally used the transmission-line taper developed by Klopfenstein [4]. The principle employed by Klopfenstein is very briefly as follows. Equation (6.5) relates the impedance function $p(x)$ to the reflection coefficient. By specifying the reflection coefficient as a continuous function, $p(x)$ can be obtained through the Fourier transform. The function

$$\rho e^{j\beta l} = \rho_0 \frac{\cos\left[\sqrt{(\beta l)^2 - A^2}\right]}{\cosh A} \tag{6.19}$$

decreases in magnitude from ρ_0 at $\omega = 0$ to $\rho = \rho_0/\cosh(A)$ when $\beta l = A$. For $\beta l > A$, the function oscillates between zero and ρ_m. It thus is a true high-pass response without an upper frequency limit. If (6.19) is substituted into (6.5), we obtain an expression from which the variation of impedance along the taper can be calculated. The tapered line is optimal because it is the shortest transformer that will match two impedances to within a fixed-ripple level. Unfortunately, the Klopfenstein taper suffers from a discontinuity at each end of the taper.

The taper proposed by Hecken [5] overcomes this difficulty but is slightly longer than the Klopfenstein taper for the same return loss. This taper is based on the approximating function

$$f(\beta) = y_0 \frac{B}{\sinh B} \cdot \frac{\sin\sqrt{(\beta l)^2 - B^2}}{\sqrt{(\beta l)^2 - B^2}} \tag{6.20}$$

Its response is not exactly equiripple, but this is of little consequence. The first ripple in the passband is the highest and is given by

$$\rho_{max} = \tanh(y_0 \epsilon_{max}) \tag{6.21}$$

where

$$\epsilon_{max} = \frac{0.21723\, B}{\sinh B} \tag{6.22}$$

$$y_0 = \tfrac{1}{2} \ln 1/k \tag{6.23}$$

The minimum taper length is found from

$$(\beta l)_{min} = \sqrt{B^2 + 6.523} \tag{6.24}$$

Figure 6.2 schematically shows the response of return loss *versus* electrical length for both the Klopfenstein and Hecken tapers. The variation of impedance levels through the coupler are shown for the even and odd modes in Figure 6.3. The Fourier transform of (6.20) is taken and, after manipulation, the characteristic impedance is calculated from

$$\ln z_C(\xi) = \tfrac{1}{2} \ln(1/k) + \tfrac{1}{2} \ln(1/k) \cdot G(B, \xi) \tag{6.25}$$

where $\xi = 2x/l$:

$$G(B, \xi) = \frac{B}{\sinh B} \int_0^{\xi} I_0\{B\sqrt{1 - \xi'^2}\} \, d\xi' \tag{6.26}$$

and I_0 is the modified Bessel function of the first kind. Values for $G(B,\xi)$ can either be calculated from (6.26) by means of a computer algorithm [6] or by making use of the tabulated values from [5] given in Table 6.1.

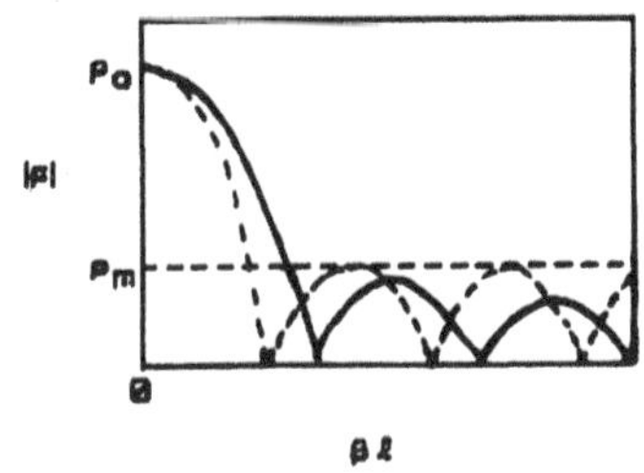

Figure 6.2 **Return loss characteristics for Klopfenstein (dotted line) and Hecken (solid line) tapers *versus* electrical length βl.**

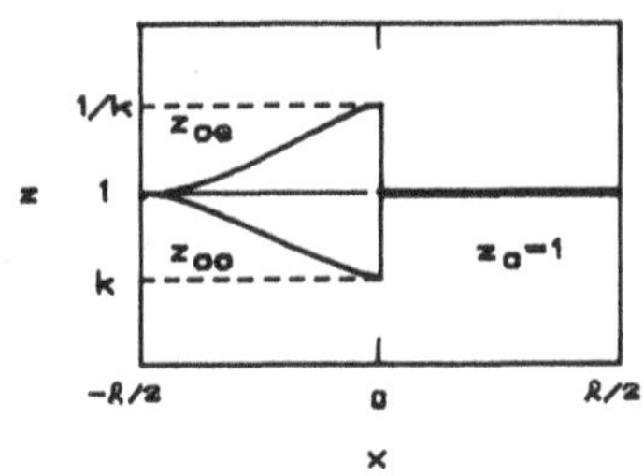

Figure 6.3 **Variation of even- and odd-mode impedances through the coupler.**

In executing a design, first obtain k and ρ_0. With the known ripple level, calculate ρ_m. Next, obtain y_0 from (6.23), and with ρ_0 known, calculate ϵ_m from (6.21). Values of $20 \log \epsilon_m$ are tabulated in Table 6.1; the corresponding values for B and $(\beta l)_{\min}$ are also tabulated.

The parameter $k(x)$ along the length of the coupled section is given by

$$k(x) = \left(\frac{z_{0o}(x)}{z_{0e}(x)}\right)^{1/2} = \frac{1}{z_{0e}(x)} \tag{6.27}$$

while the coupling coefficient can be calculated as a function of distance from

$$c(x) = \frac{Z_{0e}(x) - Z_{0o}(x)}{Z_{0e}(x) + Z_{0o}(x)} = \frac{1 - k(x)}{1 + k(x)} \tag{6.28}$$

This information can be used as inputs to LDIMS to calculate the dimensions of the coupler on a point-by-point basis; smoothing the curve connecting the various points gives the shape of the coupler.

6.2.3 Design Example

As an example, the following hybrid coupler is designed:

Nominal coupling,	c_0	$= -8.34$ dB
Ripple level,		$= \pm 0.3$ dB
Calculate,	ρ_0	$= 10 \exp(-8.34/20)$
		$= 0.3828$
	ρ_m	$= 0.0135$
	k	$= (1 - .3828)/(1 + .3828)$
		$= 0.4463$
	y_o	$= \frac{1}{2} \ln 1/.4463$
		$= 0.4033$
	ϵ_m	$= \tanh^{-1}(\rho_m)/y_0$
		$= 0.0328$
	$20 \log \epsilon_m$	$= -29.69$

The taper with $20 \log \epsilon_m = -30$ will have a slightly lower ripple level and is chosen for the design, as close enough. The coupler length at the lowest frequency is $(\beta l)_{\min} = 4.7533$. Table 6.2 shows the calculated variable values for this example.

Unfortunately, in order to analyze a coupler of this type, SYMFOUR or SYMTHREE cannot be used. Tresselt [7] discusses the conditions under which a tapered-line coupler can be analyzed; it boils down to taking a very large number of steps and performing an analysis with even and odd modes.

6.3 SYMMETRICAL 90° DIRECTIONAL COUPLERS

Previously, we mentioned that the use of a symmetrical network guarantees a 90° relative phase lead of the coupled port with respect to the transmitted port at all frequencies. In this section, we examine a method

Table 6.1 Values of the Function $G(B, \xi)$; $\xi = 2x/l$
(© 1972 IEEE, reprinted with permission, from [5])

$20 \log \epsilon_m$	−15	−20	−25	−30	−35	−40	−45	−50	−55	−60
B	1.1177	2.3204	3.2136	4.0091	4.7552	5.4710	6.1663	6.8466	7.5154	8.1752
$(\beta l)_{min}$	2.7875	3.4503	4.1046	4.7533	5.3975	6.0376	6.6741	7.3073	7.9374	8.5648
ξ					$G(B, \xi)$					
0.00	0.00000	0.00000	0.00000	0.00000	0.00000	0.00000	0.00000	0.00000	0.00000	0.00000
0.05	0.05473	0.06607	0.07514	0.08279	0.08950	0.09553	0.10106	0.10620	0.11102	0.11558
0.10	0.10939	0.13186	0.14979	0.16487	0.17806	0.18989	0.20071	0.21074	0.22012	0.22897
0.15	0.16930	0.19708	0.22345	0.24554	0.26476	0.28194	0.29757	0.31200	0.32545	0.33809
0.20	0.21819	0.26146	0.29565	0.32412	0.34874	0.37060	0.39039	0.40854	0.42535	0.44105
0.25	0.27218	0.32472	0.36593	0.39998	0.42919	0.45492	0.47802	0.49905	0.51838	0.53630
0.30	0.32580	0.38661	0.43387	0.47254	0.50539	0.53405	0.55953	0.58249	0.60341	0.62661
0.35	0.37899	0.44687	0.49907	0.54129	0.57673	0.60279	0.63415	0.65807	0.67961	0.69917
0.40	0.43166	0.50526	0.56117	0.60578	0.64272	0.67413	0.70135	0.72528	0.74654	0.76558
0.45	0.84375	0.56158	0.61986	0.66566	0.70298	0.73421	0.76085	0.78390	0.80406	0.82184
0.50	0.53519	0.61560	0.67487	0.72063	0.75726	0.78735	0.81255	0.83397	0.85237	0.86832
0.55	0.58591	0.66715	0.72597	0.77052	0.80544	0.83355	0.85660	0.87580	0.89196	0.90568
0.60	0.63586	0.71607	0.77302	0.81521	0.85754	0.87296	0.89333	0.90990	0.92352	0.93483
0.65	0.68495	0.76220	0.81587	0.85469	0.88369	0.90589	0.92321	0.93694	0.94794	0.95683
0.70	0.73314	0.80543	0.95449	0.88904	0.91412	0.93276	0.94688	0.95774	0.96618	0.97281
0.75	0.78036	0.84566	0.88886	0.91840	0.93916	0.95410	0.96504	0.97317	0.97928	0.98391
0.80	0.82655	0.88282	0.91901	0.94298	0.95924	0.97051	0.97846	0.98413	0.98824	0.99123
0.85	0.87166	0.91684	0.94505	0.96307	0.97483	0.98265	0.98940	0.99152	0.99401	0.99574
0.90	0.91564	0.94771	0.96710	0.97901	0.98646	0.99117	0.99419	0.99615	0.99743	0.99827
0.95	0.95844	0.97543	0.98534	0.99119	0.99466	0.99674	0.99800	0.99876	0.99923	0.99952
1.00	1.00000	1.00000	1.00000	1.00000	1.00000	1.00000	1.00000	1.00000	1.00000	1.00000

Table 6.2 Taper Parameters

$\lvert\xi\rvert$	$G(B, \xi)$	$z_e(+\lvert\xi\rvert)$	$z_e(-\lvert\xi\rvert)$
0.0	0.00000	1.4969	1.4969
0.1	0.16487	1.5998	1.4006
0.2	0.32412	1.7060	1.3134
0.3	0.47254	1.8112	1.2371
0.4	0.60578	1.9112	1.1724
0.5	0.72063	2.0018	1.1193
0.6	0.81521	2.0797	1.0774
0.7	0.88904	2.1426	1.0458
0.8	0.94298	2.1897	1.0233
0.9	0.97901	2.2218	1.0085
1.0	1.00000	2.2406	1.0000

for designing symmetrical couplers. Equation (5.57) states the condition for a symmetrical network: the transfer function must be of the form

$$|r_{12}|^2 = [P_n(\sin\theta)]^2$$

Kammler [8] shows that the power coupling of an multisection coupler, where the number of sections becomes very large, can be expressed as

$$\frac{|S_{12}|^2 = \Pi_N^2 \sin^2\theta}{1 + \Pi_N^2 \sin^2\theta} \tag{6.29}$$

where Π_N is found after cascade multiplication of the section transmission matrices. He discusses an optimization procedure whereby the optimum coupler performance can be obtained numerically.

Neglecting the phase term in the coupling expression (6.3) and using the fact that the network must be described by an odd function, (6.3) can be rewritten as

$$c(\omega) = \int_{-\frac{1}{2}l}^{\frac{1}{2}l} \sin 2\beta u \cdot p(u)du, \tag{6.30}$$

where

$$p(u) \doteq \frac{1}{2}\frac{d}{du} \ln z_{0e}(u) \tag{6.31}$$

The choice of $p(u)$ is once again critical, and Kammler [8] as well as Tresselt [9] discuss general philosophical aspects for the synthesis of appropriate $p(u)$-functions. The procedure that will be described here is from Tresselt [9].

6.3.1 Couplers with Periodic Zeros in $p(u)$

Figure 6.4 compares typical stepped and continuously varying couplers as to physical properties, the coupling function, and the coupling *versus* frequency response. As we will see, the continuously tapered coupler is one-quarter wavelength longer than the stepped coupler and does not exhibit a periodic response.

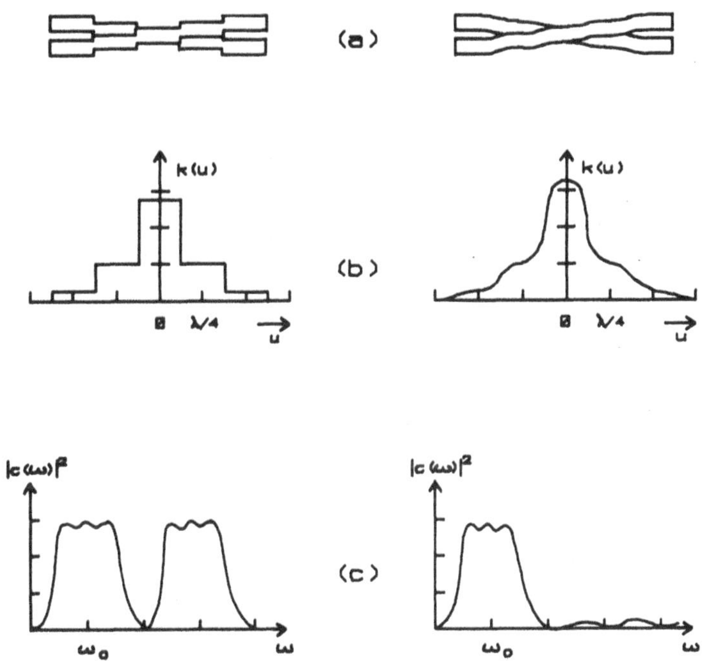

Figure 6.4 Stepped and continuously variable coupled-line couplers. The physical layout is shown in (a), with the corresponding variation of the coupling coefficient shown in (b). The stepped coupler has a periodic response, while the continuous coupler does not (c).

We have seen that the coupling function is the Fourier transform of the reflection coefficient distribution function, $p(u)$. If we choose a function of the form

$$p(u) = \frac{-1}{\pi} \frac{\sin^2 u/2}{u/2} \tag{6.32}$$

there will be unity coupling over the frequency band $0 \le \omega \le 1$, and the coupler will have to be infinitely long to accommodate all the sidelobes of $p(u)$. In order to obtain a practical coupler, the function $p(u)$ is truncated; this in turn causes a deviation from the flat unity coupling. The coupling coefficient now has nonuniform ripples, and it is necessary to distort the initial function in order to ensure a coupling with an acceptably low ripple level.

Let the truncated version of (6.32) be

$$p_T(u) = \frac{\sin^2 u/2}{u/2} \tag{6.33}$$

where the first $(N + 1)/2$ positive and negative lobes have been retained; the reason for this choice will become apparent later.

Tresselt [9] illustrates the effect of truncation by comparing the response of a truncated Fourier series of five terms ($N = 9$) with the results obtained if a function is optimized for an equiripple response. The optimized function [9] is given by

$$\begin{aligned} f(\theta) = {} & 1.2626 \sin\theta + 0.3931 \sin 3\theta + 0.2049 \sin 5\theta \\ & + 0.1170 \sin 7\theta + 0.0927 \sin 9\theta \end{aligned} \tag{6.34}$$

whereas the truncated function is

$$g(\theta) = \frac{4}{\pi} \left(\sin\theta + \tfrac{1}{3} \sin 3\theta + \tfrac{1}{5} \sin 5\theta + \tfrac{1}{7} \sin 7\theta + \tfrac{1}{9} \sin 9\theta\right) \tag{6.35}$$

The functions (6.33) and (6.34) were plotted on the same axis and shown in Figure 6.5. It is clear that the optimized response can be recovered from the truncated Fourier series by multiplying each coefficient of the latter by a weighting function:

$$w_i = c_f/c_g, \qquad i = 1 \text{ to } (2N - 1) \tag{6.36}$$

where c_f and c_g are the corresponding coefficients of the functions $f(\theta)$ and $g(\theta)$, respectively. We now need a readily obtained function, with which the truncated function can be weighted, that will ensure a good coupler performance. It turns out that one is readily at hand. Consider again the expressions for coupling, (6.30) and (6.31). In the stepped coupler, the derivative of the characteristic impedances is a series of Dirac delta functions that occur every time the impedance level changes at a step.

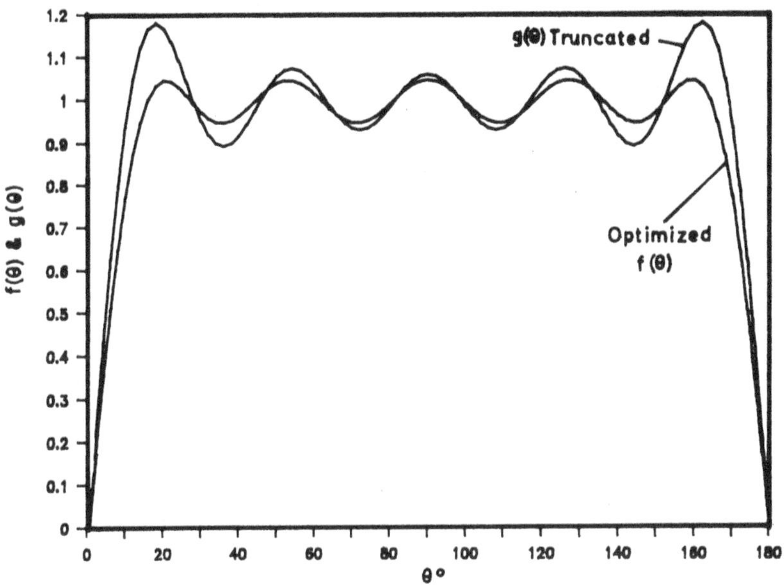

Figure 6.5 Response of a truncated and an optimized function approximating unity in the frequency interval 0 to π.

The even-mode impedances to either side of the step are Z_i and Z_{i+1}, and the area under the delta function is given by $\frac{1}{2} \ln Z_i/Z_{i+1}$, the coupling would then be given by

$$c_s(\omega) = \sum_{n=1}^{(N+1)/2} \ln \frac{Z_{((N+3)/2)-n}}{Z_{((N+1)/2)-n}} \sin [(2n - 1)\beta d/N] \tag{6.37}$$

This expression actually looks a lot more formidable than it really is, as will be illustrated by the example.

From (6.37), an expression for the coupling is obtained that is of the same form as the truncated function $g(\theta)$ shown in (6.35). Once again we compare coefficients and obtain weighting coefficients. These coefficients will ensure that, if the areas of the first $(N + 1)/2$ lobes (the ones that are retained) of the approximating function $p_\tau(u)$ are weighted by them, the same mean coupling and ripple levels as for the stepped coupler will be realized.

The weighting functions are obtained by simple comparison of the coefficients of the coupling function $c_s(\omega)$, a_i, with the coefficients of the truncated Fourier coupling function:

$$|c_c(\theta)| = \frac{4}{c\pi}\left(\sin\theta + \tfrac{1}{3}\sin 3\theta + \tfrac{1}{5}\sin 5\theta + \tfrac{1}{7}\sin 7\theta + \tfrac{1}{9}\sin 9\theta\right) \tag{6.38}$$

where $\theta = \beta l/N$, and the coupling is 20 log$c$. Let the coefficients of (6.38) be b_i. Then the weighting functions are given by

$$w_i = a_i/b_i, \qquad i = 1, 2, \ldots, (N + 1)/2 \tag{6.39}$$

The areas under the lobes of the function $p_\tau(u)$ need only be determined once and can be done directly by means of a calculator. Only the first five values are tabulated, as this is sufficient for a design of order N = 9. These values are given in Table 6.3.

Table 6.3 Areas Under the Lobes of $p_\tau(u)$

Lobe i	*Area A_i*
1	2.4377
2	0.6767
3	0.4021
4	0.2865
5	0.2226

The even mode characteristic impedance can then be obtained by integrating (6.31), where the weighted value of $p_\tau(u)$ is used:

$$\tfrac{1}{2}\ln Z_{0e}(u) = \int_o^u p_\tau(\nu)\,d\nu$$

$$= \sum_{n=Q}^{m} w_n \int_{u_n}^{u_{n-1}} \frac{\sin^2\nu/2}{\pi\nu/2}\,d\nu + w_m \int_{u_m}^{u} \frac{\sin^2\nu/2}{\pi\nu/2}\,d\nu$$

$$= \sum_{n=Q}^{m} w_n \int_{u_n}^{u_{n-1}} \frac{(1 - \cos\nu)}{\pi\nu} \, d\nu + w_m \int_{u_m}^{u} \frac{(1 - \cos\nu)}{\pi\nu} \, d\nu$$

and

$$1 \leq m \leq Q, \qquad Q = (N + 1)/2 \tag{6.40}$$

Once $Z_{0e}(u)$ is known, $k(u)$ is obtained as earlier, and dimensions can be calculated through LDIMS. Before concluding with a design example, it is interesting to note that, as the Dirac deltas on which the design is based occur in the center of each section, the coupler is always $\lambda/4$ longer than the stepped coupler.

6.3.2 Design Example

Coupling, $c = -8.34$ dB $= 0.3828$
Ripple, $= \pm 0.2$ dB
Bandwidth, $BW = 9.5$

Obtain z_i from Table 5.9,
Obtain a_i ($a_1 = 0.5054 = \ln z_5/z_4$) from (6.37),
Obtain b_i from (6.38),
Obtain w_i from (6.39).

The calculated values for these parameters are given in Table 6.4.

Table 6.4 Calculated Parameters

i	z_i	a_i	b_i	w_i
1	1.02379	0.5054	0.4874	1.0369
2	1.06452	0.1507	0.1625	0.9274
3	1.14687	0.0745	0.0975	0.7641
4	1.33341	0.0390	0.0696	0.5603
5	2.21025	0.0390	0.0542	0.4336

For this example, the lobes of $p_\tau(u)$ span the regions given in Table 6.5. The coupling is calculated by executing (6.40) over the five lobes. The cosine integrals were solved using direct integration on an HP-15C calculator (see Table 6.6). The following are sample calculations in each lobe. Let

$$\int_{u_i}^{u} \frac{\sin^2 \nu/2}{\nu/2}\, d\nu = \int_{u_i}^{u} p_\tau(\nu)\, d\nu \tag{6.41}$$

$$I_{i-1}^{i} = \int_{u_{i-1}}^{u_i} \frac{\sin^2 \nu/2}{\nu/2}\, d\nu \tag{6.42}$$

Note that, at the end of each lobe, the mean coupling in that lobe equals the mean coupling in the step of the stepped coupler. The variation of coupling *versus* distance for this example is shown in Figure 6.6.

Table 6.5 Lobe Extremities

Lobe	u_{n-1}	u_n
5	-10π	-8π
4	-8π	-6π
3	-6π	-4π
2	-4π	-2π
1	-2π	0

Table 6.6 Sample Calculation of Coupling

	u	$\int_{u_5}^{u} p_\tau(\nu)\, d\nu$	$z_{0e}(u)$	$k(u)$
Lobe 5				
$I_5^4 = 0.012$	10.0π	0.000	1.000	0.000
$cw_5/\pi = 0.053$	9.8π	0.001	1.000	0.000
$u_5 = 10\pi$	9.0π	0.108	1.012	0.012
	8.6π	0.187	1.020	0.020
	8.0π	0.223	1.024	0.024
Lobe 4				
$I_4^3 = 0.068$	7.8π	0.002	1.026	0.026
$cw_4/\pi = 0.068$	7.0π	0.137	1.043	0.042
$u_4 = 8\pi$	6.6π	0.240	1.058	0.056
	6.0π	0.286	1.065	0.063
Lobe 3				
$I_3^2 = .020$	5.8π	0.002	1.066	0.064
$cw_3/\pi = 0.093$	5.0π	0.189	1.104	0.099

Table 6.6 (cont'd.)

	u	$\int_{u_5}^{u} p_\tau(\nu)\, d\nu$	$z_{0e}(u)$	$k(u)$
$u_3 = 6\pi$	4.6π	0.335	1.135	0.126
	4.0π	0.402	1.149	0.138
Lobe 2				
$I_2^1 = .073$	3.8π	0.003	1.149	0.138
$cw_2/\pi = 0.113$	3.0π	0.304	1.230	0.204
$u_2 = 4\pi$	2.6π	0.555	1.301	0.257
	2.0π	0.667	1.338	0.283
Lobe 1				
$I_1^0 = 0.075$	1.8π	0.007	1.337	0.283
$cw_1/\pi = 0.126$	1.0π	0.789	1.629	0.453
$u_1 = 2\pi$	0.6π	1.670	2.035	0.611
	0	2.483	2.471	0.719

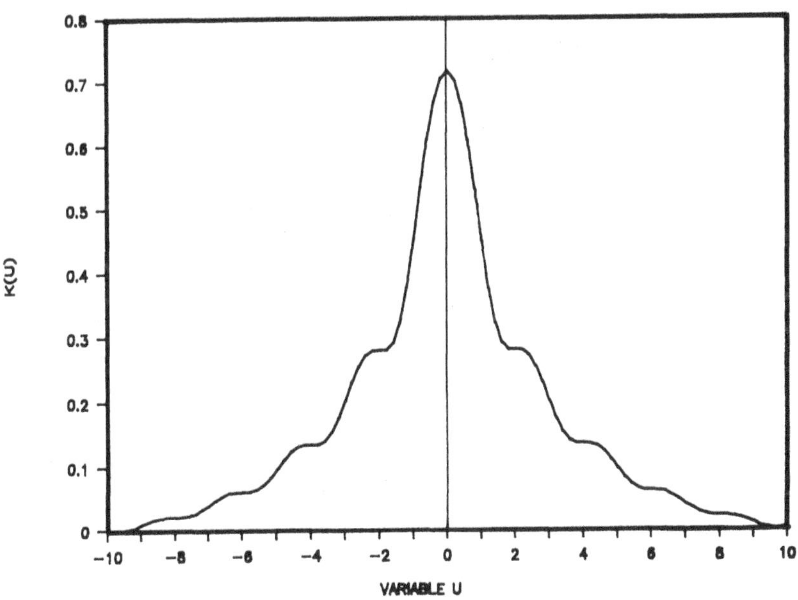

Figure 6.6 Coupling *versus* distance for a ninth order coupled-line coupler with tapered lines.

6.4 ASYMMETRIC CHEBYSCHEV HIGH-PASS COUPLERS

The final coupler of this book combines a number of the principles that have been described from time to time. In effect, it is the coupled-line realization of the equiripple stepped impedance designs described by Levy that were treated in Section 5.3.

6.4.1 Synthesis Procedure

As with the stepped asymmetric coupler, the discussion centers around the reflection coefficient of the equivalent transmission-line network, the relevant equation being (5.15):

$$\left|\Gamma\left(\frac{\cos\theta}{\cos\theta c}\right)\right|^2 = \frac{\gamma^2 - h^2 T_n^2\left(\dfrac{\cos\theta}{\cos\theta_c}\right)}{1 + \gamma^2 - h^2 T_n^2\left(\dfrac{\cos\theta}{\cos\theta_c}\right)}$$

where T_n is a Chebyschev function of the first kind of degree n and $S = j\tan\theta$. In order to retain the convention used by Arndt [10], this equation is rewritten in another form:

$$\left|\Gamma\left(\frac{\cos\theta}{\cos\theta c}\right)\right|^2 = \frac{\gamma^2 - h^2 T_{2n}\left(\dfrac{\cos\theta}{\cos\theta_c}\right)}{1 + \gamma^2 - h^2 T_{2n}\left(\dfrac{\cos\theta}{\cos\theta_c}\right)} \tag{6.43}$$

where T_{2n} is the Chebyschev function of the first kind of degree $2n$. The two forms of Γ will have the same number of ripples, and virtually identical performance, except that the maximum value of the coupling will be given by

$$|S_{11}|^2_{\min} = \frac{\gamma^2 + h^2}{1 + \gamma^2 + h^2} \tag{6.44}$$

rather than expression (5.11). The minimum coupling is given, as earlier (5.12), by

$$|S_{11}|^2_{\max} = \frac{\gamma^2 - h^2}{1 + \gamma^2 - h^2} \tag{6.45}$$

The constants γ and h are calculated from

$$\gamma^2 = \frac{1}{2}\left(\frac{C_+^2}{1 - C_+^2} + \frac{C_-^2}{1 - C_-^2}\right) \tag{6.46}$$

$$h^2 = \frac{1}{2}\left(\frac{C_+^2}{1 - C_+^2} - \frac{C_-^2}{1 - C_-^2}\right) \tag{6.47}$$

where

$$C_+ = |S_{11}| + R, \quad C_- = |S_{11}| - R \tag{6.48}$$

The coupler design parameters are shown in Figure 6.7.

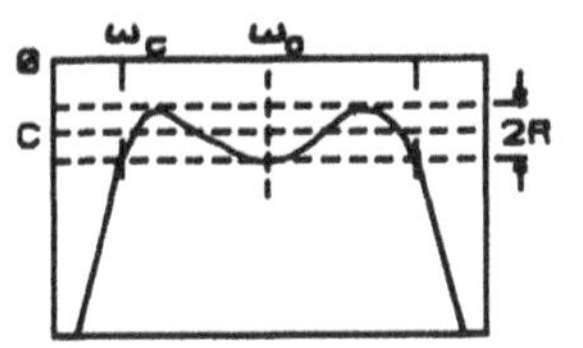

Figure 6.7 Definition of coupler parameters.

The continuously varying coupler cross section is now considered to be the limiting case of a stepped coupler, as the number of elements are allowed to become very large. The frequency variable is

$$\theta = \beta l / N \tag{6.49}$$

where l is the total length of the coupler, and N is the number of sections. For the limit, the Chebyschev function of finite order can be replaced by a cosine, so that the expression for the reflection coefficient becomes

$$|\Gamma|^2 = \frac{\gamma^2 - h^2 \cos (\sqrt{(\beta l)^2 - (\beta_c l)^2}}{1 + \gamma^2 - h^2 \cos (\sqrt{(\beta l)^2 - (\beta_c l)^2}} \tag{6.50}$$

where the cutoff variable is defined by

$$\beta_c l = \frac{1}{4\pi} \cosh^{-1} \frac{\gamma^2}{h^2} \tag{6.51}$$

The network that results from this description will be a minimum-phase network, and Arndt [10] describes the actual synthesis procedure. The ultimate result of the synthesis is the coupling factor for a large number of sections, k_i, wherein the synthesis procedure 40 sections were found to be sufficient.

6.4.2 Polynomial Representation

Rather than provide a vast number of coupling coefficients, Arndt has chosen an elegant way of presenting the design information. Using the values of the coupling coefficients, a sixth order equiripple polynomial was obtained numerically, so that the polynomial approximates the coupling function with a maximum error of 10^{-5}. The coupling is then described by

$$k(z/l) = \sum_{m=0}^{6} K_m(z/l)^m \tag{6.52}$$

Values for K_m are given in Table 6.7 from [10], for coupling levels of -3, -6, -8.34, -10, and -20 dB. The coupling is translated to even- and odd-mode impedances once again through use of equations (4.14) and (4.15):

$$z_{0e} = \sqrt{\frac{1 + k(z/l)}{1 - k(z/l)}} \tag{6.53}$$

$$z_{0o} = \sqrt{\frac{1 - k(z/l)}{1 + k(z/l)}} \tag{6.54}$$

6.4.3 Design Example

Coupling,	$c = -3$ dB
Ripple,	$R = \pm 0.17$ dB
Cutoff frequency,	$f_c = 1$ GHz
From Table 6.6,	$l/\lambda_c = 0.256$
Material choice,	$\epsilon_r = 2.32$
Calculate	$\lambda_c = 300/\sqrt{2.32}$ mm
	$l = 50.44$ mm

Table 6.7 Coefficients of the Coupling Factor Polynomial (© IEEE 1970, reprinted with permission, from [10]; additional values available in [10])

R dB	*0.047*	*0.172*	*0.341*	*0.506*	*0.749*
Coupling: −3 dB					
l/λ_c	0.366	0.256	0.201	0.169	0.236
K_0	0.9426	0.9433	0.9436	0.9441	0.9455
K_1	−0.5125	−0.4430	−0.3801	−0.3369	−0.2880
K_2	−1.4506	−0.7860	−0.6446	−0.5655	−0.4720
K_3	−3.8273	−2.7976	−1.6254	−1.0056	−0.5231
K_4	13.3169	7.0502	3.5327	1.9161	0.7889
K_5	−12.2045	−5.2493	−2.1201	−0.8610	−0.1260
K_6	3.7053	1.3389	0.4068	0.0805	−0.718
Coupling: −6 dB					
l/λ_c	0.398	0.288	0.233	0.210	0.160
K_0	0.7997	0.8000	0.8003	0.8006	0.8014
K_1	−0.9409	−0.8007	−0.7135	0.6547	0.5882
K_2	−1.5529	−0.9448	−0.6955	−0.5668	−0.4547
K_3	2.2715	0.7188	0.3054	0.1565	0.0920
K_4	1.0843	1.6392	1.3332	1.0693	0.7214
K_5	−2.7446	−2.0684	−1.3916	−1.0066	−0.6055
K_6	1.0901	0.6829	0.4152	0.2816	0.1533
Coupling: −8.34 dB					
l/λ_c	0.409	0.299	0.243	0.211	0.179
K_0	0.6687	0.6688	0.6690	0.6691	0.6695
K_1	−1.0507	−0.8890	−0.7923	−0.7307	−0.6607
K_2	−0.7719	−0.4869	−0.3625	−0.2926	−0.2354
K_3	2.3164	1.1058	0.6730	0.4642	0.3155
K_4	−1.0692	−0.0433	0.1538	0.1973	0.1822
K_5	−0.3929	−0.5831	−0.4768	−0.3768	−0.2748
K_6	0.3044	0.2459	0.1730	0.1235	0.0847
Coupling: −10 dB					
l/λ_c	0.413	0.303	0.248	0.214	0.183
K_0	0.5749	0.5750	0.5750	0.5751	0.5754
K_1	−1.0252	−0.8671	−0.7739	−0.7132	−0.6459
K_2	−0.3166	−0.1906	−0.1412	−0.1143	−0.0953
K_3	1.7708	0.8900	0.5824	0.4255	0.3117

Table 6.7 Coefficients of the Coupling Factor Polynomial (© IEEE 1970, reprinted with permission, from [10]; additional values available in [10])

R dB	*0.047*	*0.172*	*0.341*	*0.506*	*0.749*
K_4	−1.1930	−0.3287	−0.1173	−0.0378	−0.0127
K_5	0.0893	−0.1788	−0.1748	−0.1530	−0.1119
K_6	0.1034	0.1054	0.0778	0.0598	0.0420
Coupling: −20 dB					
l/λ_c	0.421	0.310	0.255	0.223	0.191
K_0	0.1980	0.1981	0.1981	0.1980	0.1980
K_1	−0.4440	−0.3768	−0.3373	−0.3107	−0.2825
K_2	0.2070	0.1594	0.1309	0.1063	0.0887
K_3	0.2014	0.0786	0.0414	0.0425	0.0272
K_4	−0.1994	−0.0456	−0.0140	−0.0133	−0.0028
K_5	0.0210	−0.0236	−0.0175	−0.0209	−0.0183
K_6	0.0172	0.0140	0.0065	0.0103	0.0080

From Table 6.6,

$K_0 = 0.9433 \qquad K_1 = -0.4430$

$K_2 = -0.7860 \qquad K_3 = -2.7976$

$K_4 = 7.0502 \qquad K_5 = -5.2493$

$K_6 = 1.3389$

Setting $z/l = \zeta$,

$$k(\zeta) = \mathbf{0.9433 - 0.4430\zeta - 0.7860\zeta^2 - 2.7976\zeta^3 + 7.0502\zeta^4 - 5.2493\zeta^5 + 1.3389\zeta^6}$$

Figure 6.8 shows the variation of coupling *versus* distance for the coupler.

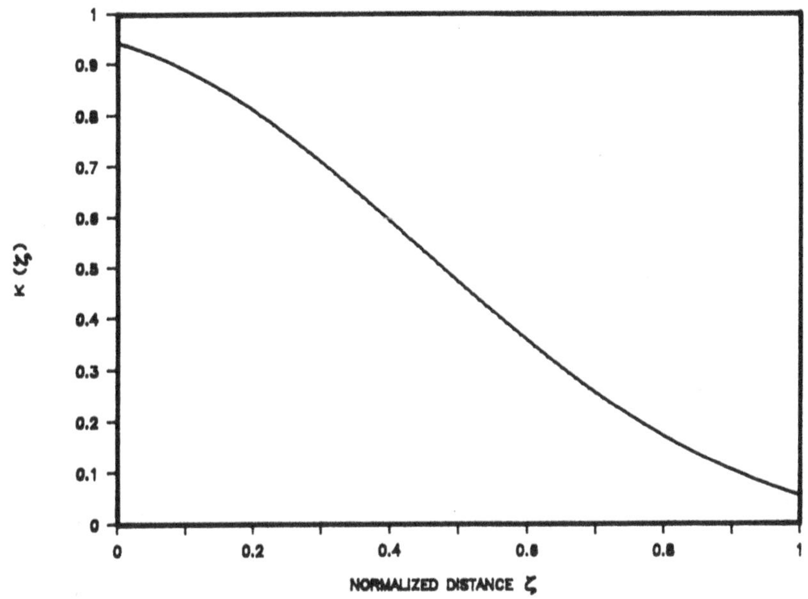

Figure 6.8 **Variation of coupling with distance for an unsymmetric high-pass directional coupler.**

REFERENCES

[1] C. B. Sharpe, "An Equivalence Principle for Nonuniform Transmission-Line Directional Couplers," *IEEE Trans. Microwave Theory Tech.*, Vol. MTT-15, July 1967, pp. 398–405.

[2] C. B. Sharpe, "An Alternative Derivation of Orlov's Synthesis Formula for Non-Uniform Lines," *IEE Proc.*, Vol. 109, Monograph 483E, November 1961, pp. 226–229.

[3] R. H. DuHamel and M. E. Armstrong, "The Tapered-Line Magic-T," *Abstracts of the Fifteenth Annual Symposium on the USAF Antenna Research Program*, Monticello, IL, October 12–14, 1965.

[4] R. W. Klopfenstein, "A Transmission Line Taper of Improved Design," *IRE Proc.*, Vol. 44, January 1956, pp. 31–35.

[5] R. P. Hecken, "A Near-Optimum Matching Section without Discontinuities," *IEEE Trans. Microwave Theory Tech.*, Vol. MTT-20, November 1972, pp. 734–739.

[6] J. H. Cloete, "Rapid Computation of the Hecken Impedance Function," *IEEE Trans. Microwave Theory Tech.*, Vol. MTT-25, May 1977, p. 440.

[7] C. P. Tresselt, "Design and Computed Theoretical Performance of Three Classes of Equal-Ripple Nonuniform Line Couplers," *IEEE Trans. Microwave Theory Tech.*, Vol. MTT-17, April 1969, pp. 218–230.

[8] D. W. Kammler, "The Design of Discrete N-Section and Continuously Tapered Symmetrical Microwave TEM Directional Couplers," *IEEE Trans. Microwave Theory Tech.*, Vol. MTT-17, August 1969, pp. 577–590.

[9] C. P. Tresselt, "The Design and Construction of Broadband, High-Directivity, 90-Degree Couplers Using Nonuniform Line Techniques," *IEEE Trans. Microwave Theory Tech.*, Vol. MTT-14, December 1966, pp. 647–656.

[10] F. Arndt, "Tables for Asymmetric Chebyschev High-Pass TEM-Mode Directional Couplers," *IEEE Trans. Microwave Theory Tech.*, Vol. MTT-18, September 1970, pp. 633–638.

Appendix A
Network Analysis Programs

A1 INTRODUCTION

This appendix describes the operation and function of the computer programs SYMFOUR.BAS and SYMTHREE.BAS; they run in GWBASIC. SYMFOUR is used to analyze the response of a symmetrical four-port network consisting of reactive elements, capacitive elements, unit elements, and resistors, with the two preconditions that it must have four ports and that the analysis is carried out with respect to two of the ports on either side of a line of symmetry.

The program SYMTHREE performs a similar function, but it analyzes the properties of a three-port network with symmetry on either side of two of the ports and an axis of symmetry running through the third port, as is often found in power dividers.

Both programs have a graphics output; it is essential that the program run in the correct graphics environment for a specific computer. Determine the relevant statement and correct lines 2360 in both programs if necessary (the SCREEN statement).

A2 PROGRAM SYMFOUR

A2.1 Method of Calculation

SYMFOUR makes use of the eqs. (1.45)–(1.48) to calculate the scattering parameters of a symmetrical four-port network, as seen from one of the ports. The port, in this case designated port 1, must lie to one side of the axis of symmetry, and all information is entered into the program with respect to that port.

The program calculates the input admittance by means of a ladder network calculation. It starts off with an admittance of 0 + j0 and adds the admittances in order as indicated by their type. First, the type of element is identified, and whether it is connected in series or in parallel is determined. If in parallel, the admittance is added; otherwise, the inverse is calculated and added. A subroutine transforms the admittances through a unit element where necessary.

Should there be a branch in the circuit, the position of the connection of the branch is indicated. When that point is reached, the admittance is stored and the input admittance of the branch is calculated by starting from its terminated end. When the driving point of the branch is reached, the way in which the branch is connected would have been indicated (connected in series or parallel). The admittances are then inverted, if necessary, and added. Finally, the input reflection coefficient is calculated.

Starting with an assumed voltage of 1V at the terminated port, the voltage is calculated all along the network as the impedance calculation proceeds. The voltage obtained at the input terminals is then used with the starting voltage to determine the transmission coefficient.

Once both coefficients have been calculated for the even-mode network, the process is repeated for the odd-mode network. The scattering parameters are calculated from these variables.

A2.2 Preparation

Prior to running the program, the network has to be prepared for analysis. The following steps are taken:

1. Terminate all ports.
2. Divide the network into the even- and odd-mode networks. For the even mode, a magnetic conductor is introduced all along the axis of symmetry. Any element not crossing this line remains invariant; all elements crossing the line are open circuited at the point of crossing, as was done for the branch-line coupler in Figure 2.5(a) and 2.5(c).
3. For the odd-mode network, an electric conductor is introduced along the axis of symmetry, and all crossings are short circuited, as in Figure 2.5(b) and 2.5(d).
4. The element values are now designated and coded, as

   [TYPE,VALUE,FREQ]

   where the TYPE is specified as follows:

PR = Parallel Resistor
PC = Parallel Capacitor
PI = Parallel Inductor
SR = Series Resistor
SC = Series Capacitor
SL = Series Inductor
U = Cascade Unit Element
B = Start of a Branch in the Circuit
PB = End of the Branch, connect in parallel
SB = End of the Branch, connect in series

The VALUE is the admittance of the element for all elements, normalized to a port admittance of 1. FREQ is the resonant frequency of the element, and equals 1 for a quarterwave section, ½ for a halfwave section, and 2 if the section is one-eighth of a wavelength long at the center frequency of the network.

A2.3 Running

After loading the program and the RUN command, the program will respond,

```
                               *************
Fourport Analysis Program      *  SYMFOUR  *
                               *************
Element Codes:

      N  = Order of Network
      SR = Series Resistor
      PR = Shunt Resistor
      U  = Unit Element
      SI = Series Inductor
      PI = Shunt Inductor
      SC = Series Capacitor
      PC = Shunt Capacitor
      B  = Start of a Branch Circuit
      SB = End of Branch, connected in Series
      PB = End of Branch, connected in Parallel

*********************************************

J.A.G. Malherbe, University of Pretoria, 1987

*********************************************

ORDER OF NETWORK, M=?
```

to which the operator responds with the maximum number of elements in the odd or even mode network, including port terminations. The computer then responds with

```
****************************************
NOW TYPE IN THE ELEMENT VALUES, FOLLOWED BY THEIR
ADMITTANCE VALUES, AND RESONANT FREQUENCIES
****************************************
EVEN MODE NETWORK:
****************************************
```

The element values are now entered in the format

```
[TYPE,VALUE,FREQ]RETURN
```

as described earlier; after each element the prompt will reappear, until all the even-mode information has been entered, after which

```
****************************************
END OF EVEN MODE NETWORK
****************************************
****************************************
ODD MODE NETWORK:
****************************************
```

The process is now repeated for the odd-mode network, at the end of which, the message will appear

```
****************************************
END OF ODD MODE NETWORK
****************************************
CALCULATING .......
```

The next reaction comes after the grid for display of the chosen *S*-parameters has appeared on the screen. The computer prints

```
X-RANGE  =  50  TO  150
Y-RANGE  =  -50  TO 0
WHICH PARAMETER WOULD YOU LIKE TO SEE?
```

To which you reply with the relevant S-parameter; for instance, 2 ENTER for S_{22}, *et cetera* (1, 2, 3, or 4). The X-range is given as a percentage of the center frequency; that is, in the example, the display will be of 0.5 to 1.5 times the center frequency. The Y-range is in decibels.

For very wideband cases, it will be desirable to modify the displayed bandwidth. This may be done by listing and changing line 590 in the program, which reads

```
590 XBEGIN=50:XEINDE=150:XSTAP=3
```

In this line, XBEGIN and XEINDE are the start and stop frequencies, respectively, and XSTAP, the step size. Allowable values are 0 < XBEGIN < 100; 100 < XEINDE < 200. XSTAP may be decreased if a finer detail is wanted in the displayed curves or increased (to a value of, say, 10) if a more rapid calculation with less detail is desired. The desired response is now plotted. After the plot, the prompt appears,

```
TO REVIEW WITH NEW SCALE,
TYPE # OF dB, OTHERWISE 0?
```

If the scale is changed, the grid will be redrawn. The option

```
CLEAR SCREEN?
```

is used as convenient. It allows the display of all the S-parameters on the same plot or whatever combinations are convenient. The program continues to return to the point where you can exercise the option to review with a new scale.

A2.4 Analysis Examples

A2.4.1 Branch-line Hybrid

The branch-line hybrid analysis example refers to the network discussed in Section 2.3.1. The layout of the −3 dB branch-line coupler repeated in Figure A1(a), while Figures A1(b) and A1(c) show the even- and odd-mode bisections of the network, with the element values as normalized admittances. The equivalent circuits for the two networks are shown in A1(d) and A1(e).

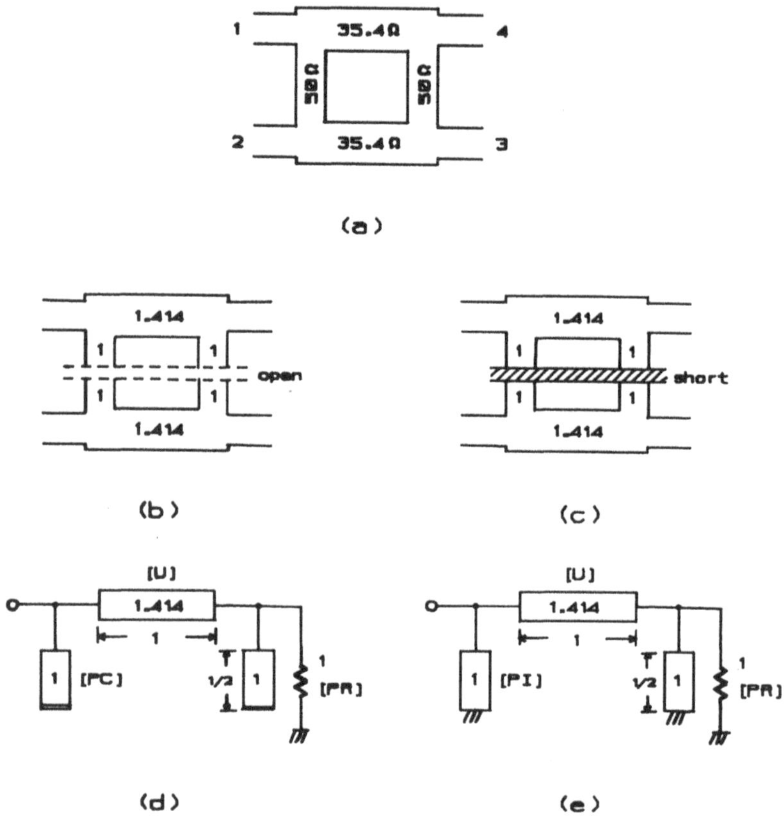

Figure A1 **(a) Branch-line coupler for analysis with even- (b) and odd-mode (c) networks. The even- and odd-mode equivalent circuits are shown in (d) and (e).**

The input to the computer program SYMFOUR is then as follows:

Even mode:

PR, 1, 1	(Parallel Resistor, 1S, dummy)
PC, 1, 2	(Parallel Capacitor, 1S, $2 \cdot f_0$)
U, 1.414, 1	(Unit Element, 1.414S, $1 \cdot f_0$)
PC, 1, 2	(Parallel Capacitor, 1S, $2 \cdot f_0$)

Odd mode:

PR, 1, 1	(Parallel Resistor, 1S, dummy)
PI, 1, 2	(Parallel Inductor, 1S, $2 \cdot f_0$)
U, 1.414, 1	(Unit Element, 1.414S, $1 \cdot f_0$)
PI, 1, 2	(Parallel Inductor, 1S, $2 \cdot f_0$)

Note the use of a dummy variable in the input data for the resistor. Note also that the termination must always be given as a parallel element, as the initial value that the program starts with sets the real and imaginary values to zero; if the first element is defined as a series element, a singularity occurs.

Figure A2 shows the display during startup and entering of the data, while the branch-line hybrid performance shown in Figure 2.7 is repeated in Figure A3.

```
RUN
ORDER OF NETWORK, M=? 4
**************************
NOW TYPE IN THE ELEMENT VALUES, FOLLOWED BY THEIR
ADMITTANCE VALUES, AND RESONANT FREQUENCIES
**************************
EVEN MODE NETWORK:
**************************
? PR,1,1
? PC,1,2
? U,1.414,1
? PC,1,2
**************************
END OF EVEN MODE NETWORK
**************************
**************************
ODD MODE NETWORK:
**************************
? PR,1,1
? PI,1,2
? U,1.414,1
? PI,1,2
**************************
END OF ODD MODE NETWORK
**************************
CALCULATING .......
```

Figure A2 Computer screen display for −3 dB branch-line coupler analysis.

A2.4.2 Optimized Hybrid Ring

The optimized hybrid ring is an example of a branch in the analysis of a network and, as such, is informative. The example analyzed here is discussed in Section 2.4.4, and the physical layout shown in Figure 2.17 is repeated here in Figure A4.

The even-mode network is obtained by open circuiting along the axis of symmetry, shown in Figure A5(a). The odd-mode network, obtained through shorting along the same line, is shown in Figure A5(b). In each case, the elements and their designation are shown.

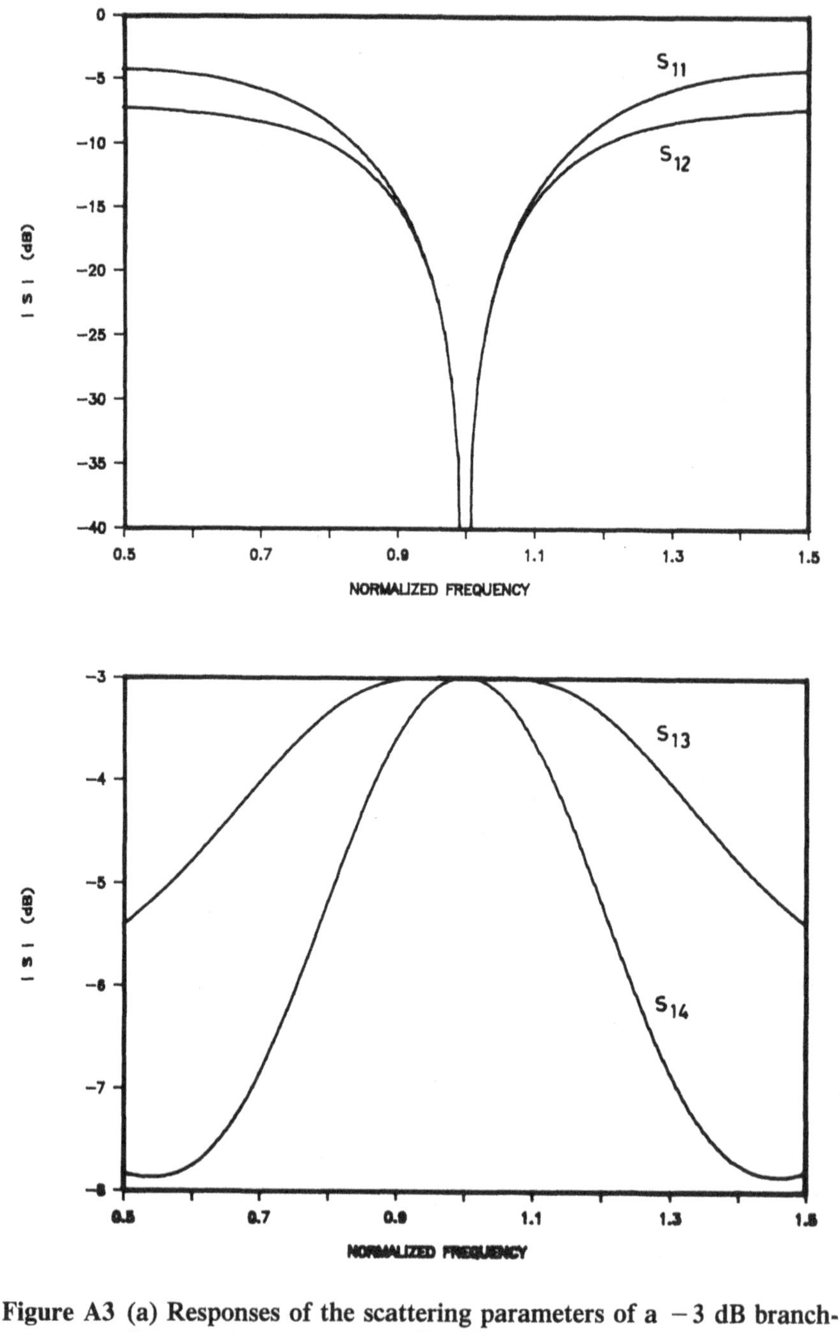

Figure A3 (a) Responses of the scattering parameters of a −3 dB branch-line coupler *versus* normalized frequency. (b) Expanded view of the coupling and transmission.

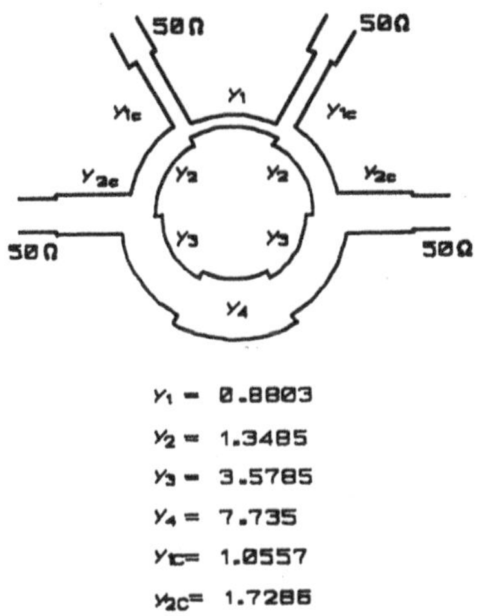

Figure A4 Optimized hybrid ring coupler.

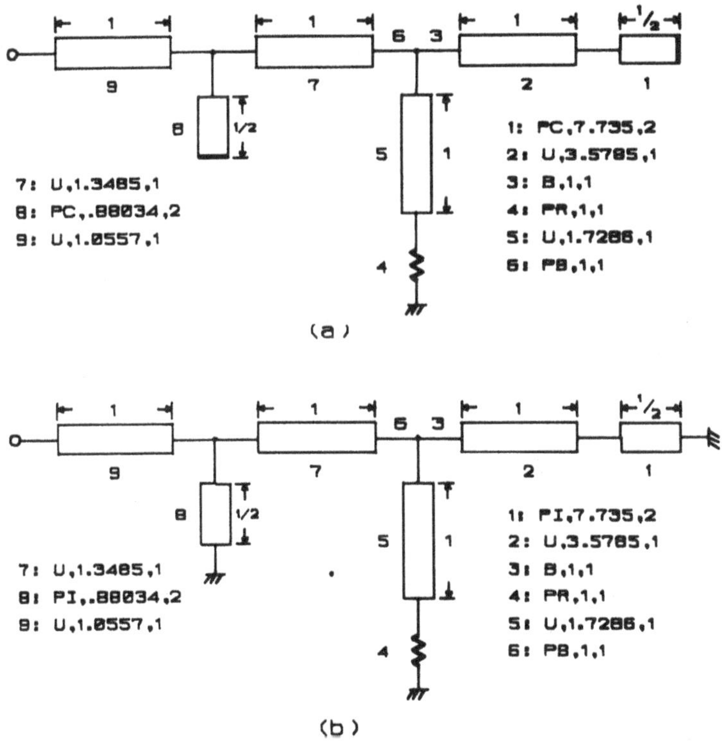

Figure A5 Even- and odd-mode networks for the example of an optimized wideband ring coupler.

The branch occurs after element number 2 has been entered, and the instruction to branch is itself entered as element number 3, with two dummy variables. The branch, consisting of elements 4 and 5, is entered next, and element number 6 is the instruction to terminate the branching and specifies that the branch is a parallel branch.

The entry of the data is shown in Figure A6, and the scattering parameter responses for the network are shown in Figure A7. (The same data is shown in Figure 2.18.)

```
RUN
ORDER OF NETWORK, M=? 9
**************************
NOW TYPE IN THE ELEMENT VALUES, FOLLOWED BY THEIR
ADMITTANCE VALUES, AND RESONANT FREQUENCIES
**************************
EVEN MODE NETWORK:
**************************
? PC,7.735,2
? U,3.5785,1
? B,1,1
? PR,1,1
? U,1.7286,1
? PB,1,1
? U,1.3485,1
? PC,0.88034,2
? U,1.05570,1
**************************
END OF EVEN MODE NETWORK
**************************
**************************
ODD MODE NETWORK:
**************************
? PI,7.735,2
? U,3.5785,1
? B,1,1
? PR,1,1
? U,1.7286,1
? PB,1,1
? U,1.3485,1
? PI,0.88034,2
? U,1.05570,1
**************************
END OF ODD MODE NETWORK
**************************
CALCULATING .......
```

Figure A6 Computer screen display for optimized ring coupler analysis.

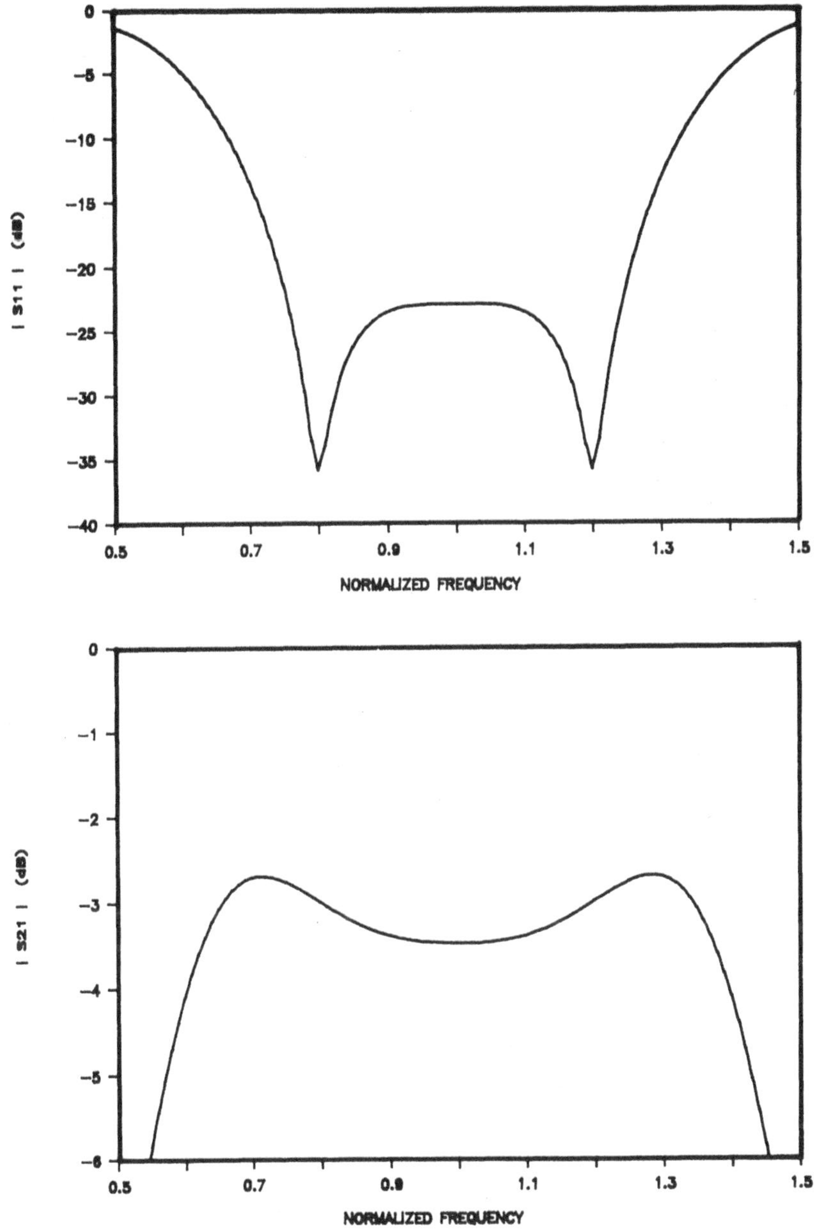

Figure A7 (a) Responses of the scattering parameters of an optimized −3 dB hybrid ring *versus* normalized frequency. (b) Expanded view of the coupling.

A2.4.3 Wideband Branchline Coupler

Figure A8 is a copy of the screen display when the wideband branch-line coupler shown in Figure A9 is analyzed. The odd and even networks each consist of 6 elements, including the end resistive termination. The coupler response is shown in Figure A10.

```
RUN
ORDER OF NETWORK, M=? 6
*****************************
NOW TYPE IN THE ELEMENT TYPES, FOLLOWED BY THEIR
ADMITTANCE VALUES, AND RESONANT FREQUENCIES
*****************************
EVEN MODE NETWORK:
*****************************
? PR,1,1
? PC,.4545,2
? U,1.0792,1
? PC,.7,2
? U,1.0792,1
? PC,.4545,2
*****************************
END OF EVEN MODE NETWORK
*****************************
ODD MODE NETWORK:
*****************************
? PR,1,1
? PI,.4545,2
? U,1.0792,1
? PI,.7,2
? U,1.0792,1
? PI,.4545,2
*****************************
END OF ODD MODE NETWORK
*****************************
CALCULATING ........
```

Figure A8 Computer display while the element values of a wideband branch-line coupler are being entered.

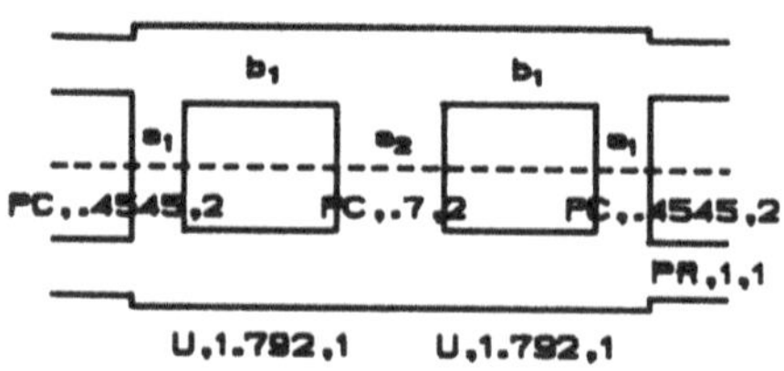

Figure A9 Diagrammatic representation of the branch-line coupler 3-14 being analyzed.

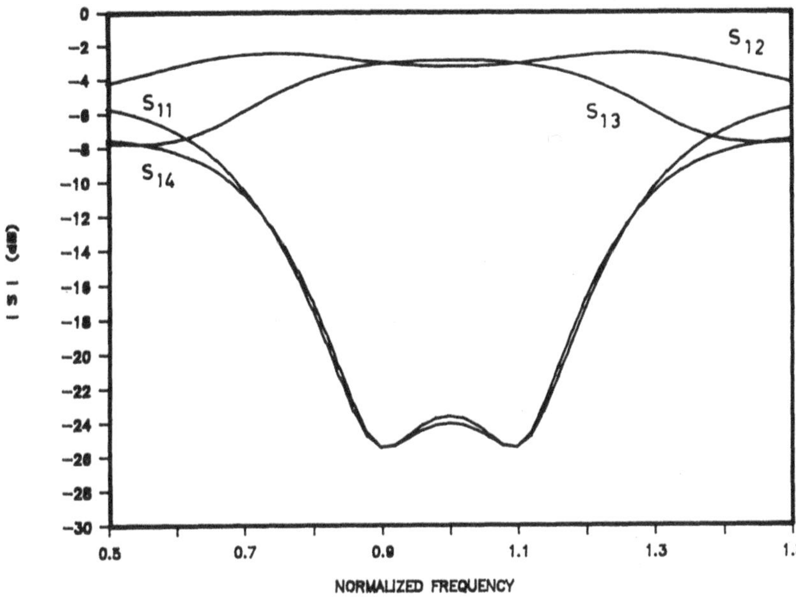

Figure A10 Calculated scattering parameters of a third order branch-line coupler.

A2.4.4 Coupled-Line Section

Coupled-line sections can be analyzed by means of SYMFOUR as well. The coupled-line section is replaced by its even- and odd-mode admittances and treated as if it were a normal connected-line network. Figure A11 shows the pair of coupled lines that is treated as an example in Section 4.2.5.

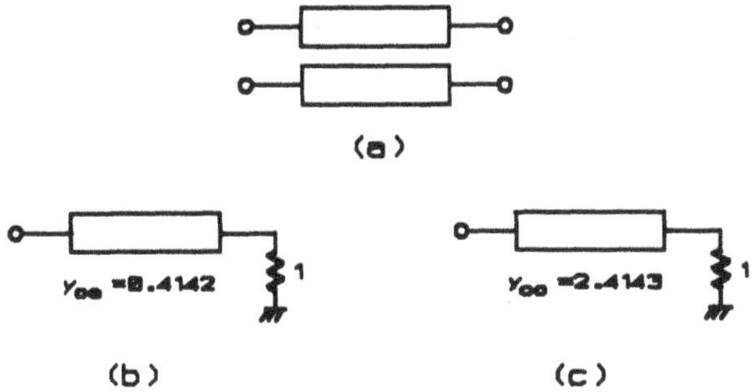

Figure A11 Coupled-line directional coupler (a), with even-mode (b), and odd-mode network (c).

The design values from the example were

$$Z_{0e} = 120.71\ \Omega, \qquad y_{0e} = 0.4142$$

$$Z_{0o} = 20.71\ \Omega, \qquad y_{0o} = 2.4143$$

and the input to SYMFOUR is shown in Figure A12. The response obtained is shown in Figure A13 (also in Figure 4.2 in Chapter 4).

```
RUN
ORDER OF NETWORK, M=? 2
**************************
NOW TYPE IN THE ELEMENT VALUES, FOLLOWED BY THEIR
ADMITTANCE VALUES, AND RESONANT FREQUENCIES
**************************
EVEN MODE NETWORK:
**************************
? PR,1,1
? U,2.414,1
**************************
END OF EVEN MODE NETWORK
**************************
**************************
ODD MODE NETWORK:
**************************
? PR,1,1
? U,.414,1
**************************
END OF ODD MODE NETWORK
**************************
CALCULATING .......
```

Figure A12 Computer screen display for coupled-line directional coupler analysis in the example.

A3 PROGRAM SYMTHREE

A3.1 Method of Calculation

SYMTHREE is a modified version of SYMFOUR, which is used specifically for the analysis of three-port networks with two ports on either side of the line of symmetry and the third port on the line of symmetry, as is the case with power dividers. The comments on scales and changes in the horizontal display, *et cetera* that were applicable to SYMFOUR apply in this case as well, to the extent that the same line numbers are used.

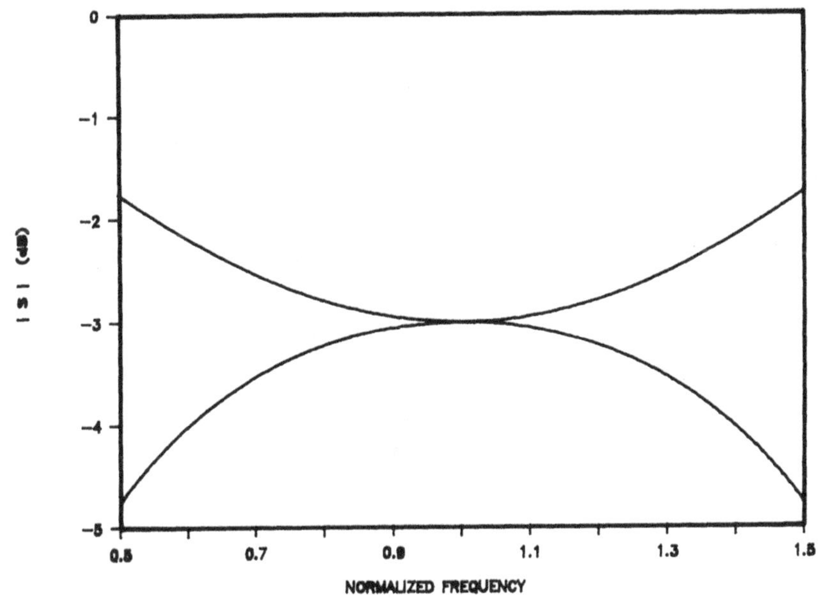

Figure A13 **Computer screen display and calculated responses of *S*-parameters *versus* frequency for the design example of a −3 dB coupled-line coupler.**

The reference port in this case is one of the two output ports; the input port reflection coefficient is calculated from the knowledge that $S_{11} = \Gamma_e$ (answer 1 to the *S*-parameter choice), and the forward transmission coefficient S_{21} or S_{31} is obtained from the unitary condition (answer 2 for S_{12}). The reflection coefficient and isolation at the output end is determined in the same fashion as in program SYMFOUR.

A3.2 Preparation and Running

The network is prepared in the same way as earlier, except that there is now a port on the axis of symmetry. It should be noted that there no longer is a one-to-one correspondence of elements between the even- and odd-mode networks. As a matter of fact, for any higher order power divider (say, of order N), there will be $N + 1$ unit elements in the even-mode network. (N unit elements do not cross the line of symmetry, and one terminates on the line of symmetry. The other resistors are open circuited and thus disappear from the even-mode network.)

For the odd-mode network there will be $N - 1$ unit elements, one inductor (the last unit element is shorted at the junction and becomes an inductor) and up to N shunt resistors connected to the grounded center line. The difference in the number of elements in the odd- and even-mode networks must be made up by adding, preferably at the output of the even-mode network, the necessary number of dummy elements, such as [PC,0,1].

For compensated power dividers, the input port termination is fed through a unit element. Instead of replacing the termination by an effective termination of $y/2$, the even-mode network has an additional unit element of admittance one-half the unit element admittance. In this case, of course, the odd-mode network again has one element less. This is illustrated in Figure 2.19. The procedure is now illustrated by means of an example.

A3.3 Two-Section Power Divider

A three-port divider, together with its even- and odd-mode networks is shown in Figure A14. (This network is also shown in Figure 3.10.) The input and operating procedure for running the program SYMTHREE is illustrated in Figure A15, and its calculated performance is shown in Figure A16. Note that the maximum number of elements needed is the same as for the odd-mode network (viz., four), so a dummy admittance of zero is introduced to the one end of the network.

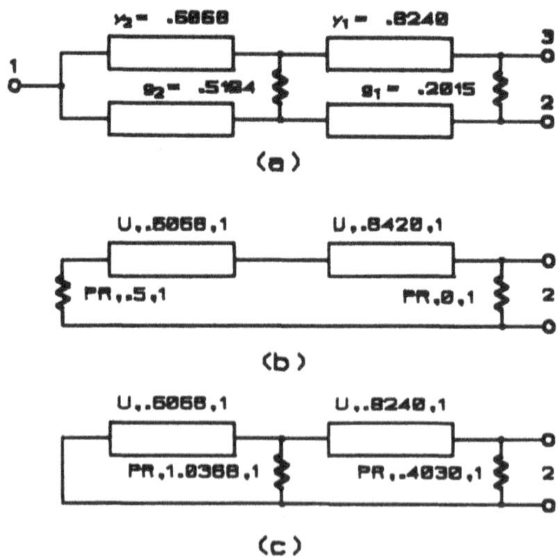

Figure A14 Example of a two-section power divider with 0.6 relative bandwidth (a). The even-mode network is shown in (b), where the right admittance is a dummy variable. The odd-mode network is shown in (c).

```
RUN
ORDER OF NETWORK, M=? 4
**************************
NOW TYPE IN THE ELEMENT VALUES, FOLLOWED BY THEIR
ADMITTANCE VALUES, AND RESONANT FREQUENCIES
**************************
EVEN MODE NETWORK:
**************************
? PR,.5,1
? U,.6068,1
? U,.8240,1
? PR,0,2
**************************
END OF EVEN MODE NETWORK
**************************
**************************
ODD MODE NETWORK:
**************************
? PI,.6068,1
? PR,1.0368,1
? U,.8240,1
? PR,.4030,1
**************************
END OF ODD MODE NETWORK
**************************
CALCULATING .......
```

Figure A15 Format of data entry to program SYMTHREE for the analysis of a two-section power divider.

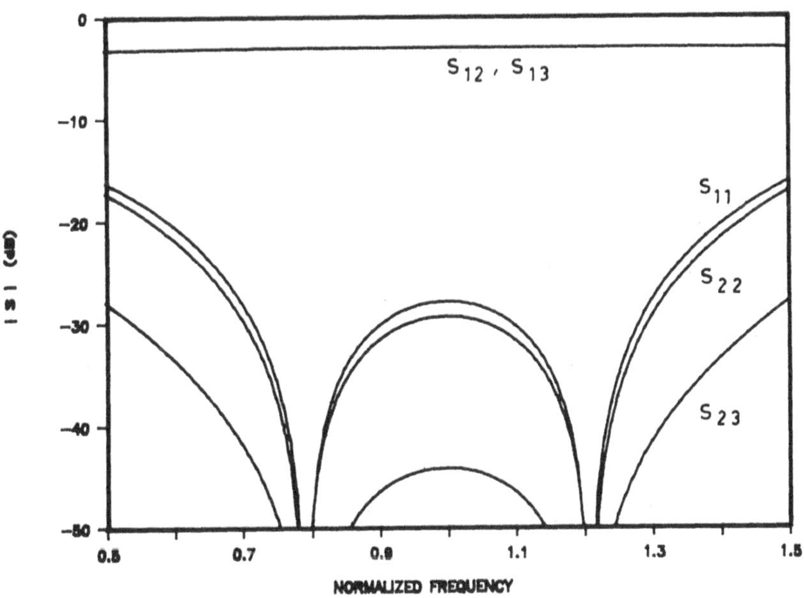

Figure A16 Calculated performance of two-section power divider with equi-ripple even-mode and odd-mode reflection coefficients.

A4 Program Listings

A4.1 SYMFOUR

```
10 REM **************************************
20 REM *                                    *
30 REM *             SYMFOUR                *
40 REM *                                    *
50 REM **************************************
60 REM
70 PI=3.1415926535#
80 REM Version # 4.04     Date:  10-04-1988
90 REM Program analyzes symmetrical ladder networks
100 REM by means of odd and even mode networks.
110 CLS
120 REM  "********************************************"
130 PRINT "                          *************
140 PRINT "Fourport Analysis Program  *  SYMFOUR  *"
150 PRINT "                          *************"
160 PRINT "Element Codes:"
170 PRINT " "
180 PRINT "      N = Order of Network"
190 PRINT "     SR = Series Resistor"
200 PRINT "     PR = Shunt Resistor"
210 PRINT "      U = Unit Element"
220 PRINT "     SI = Series Inductor"
230 PRINT "     PI = Shunt Inductor"
240 PRINT "     SC = Series Capacitor"
250 PRINT "     PC = Shunt Capacitor"
260 PRINT "      B = Start of a Branch Circuit"
270 PRINT "     SB = End of Branch, connected in Series"
280 PRINT "     PB = End of Branch, connected in Parallel"
290 PRINT " "
300 PRINT "*********************************************"
310 PRINT " "
320 PRINT "J.A.G. Malherbe, University of Pretoria, 1987"
330 PRINT " "
340 PRINT "*********************************************"
350 REM
360 REM START OF DATA INPUT
370 DIM RYE(200),XYE(200),RYO(200),XYO(200),RGE(200),XGE(200)
380 DIM RGO(200),XGO(200),VRE(200),VXE(200),VRO(200),VXO(200)
390 INPUT "ORDER OF NETWORK, M=";M
400 PRINT "*****************************"
410 PRINT "NOW TYPE IN THE ELEMENT TYPES, FOLLOWED BY THEIR "
420 PRINT "ADMITTANCE VALUES, AND RESONANT FREQUENCIES"
```

```
430 PRINT "******************************"
440 PRINT "EVEN MODE NETWORK:
450 PRINT "******************************
460 FOR I=1 TO M:INPUT TYE$(I),VAE(I),EFR(I):NEXT I
470 PRINT "******************************"
480 PRINT "END OF EVEN MODE NETWORK"
490 PRINT "******************************"
500 REM
510 PRINT "******************************"
520 PRINT "ODD MODE NETWORK:"
530 PRINT "******************************"
540 FOR I=1 TO M:INPUT TYO$(I),VAO(I),OFR(I):NEXT I
550 PRINT "******************************"
560 PRINT "END OF ODD MODE NETWORK"
570 PRINT "******************************"
580 PRINT "CALCULATING ........"
590 XBEGIN=50:XEINDE=150:XSTAP=3
600 FOR J=XBEGIN TO XEINDE STEP XSTAP:F=.01*J+1E-13
610 REM CALCULATION OF YE NOW FOLLOWS
620 REM
630 R=0:X=0:VR=1:VX=0
640 FOR I=1 TO M
650 IF TYE$(I)="SR" OR TYE$(I)="sr" THEN GOTO 710 ELSE IF
    TYE$(I)="PR" OR TYE$(I)="pr" THEN GOTO 760
660 IF TYE$(I)="U" OR TYE$(I)="u" THEN GOTO 790 ELSE IF
    TYE$(I)="SI" OR TYE$(I)="si" THEN GOTO 830
670 IF TYE$(I)="PI" OR TYE$(I)="pi" THEN GOTO 880 ELSE IF
    TYE$(I)="SC" OR TYE$(I)="sc" THEN GOTO 910
680 IF TYE$(I)="PC" OR TYE$(I)="pc" THEN GOTO 960 ELSE IF
    TYE$(I)="B" OR TYE$(I)="b" THEN GOTO 990
690 IF TYE$(I)="SB" OR TYE$(I)="sb" THEN GOTO 1020 ELSE IF
    TYE$(I)="PB" OR TYE$(I)="pb" THEN GOTO 1070
700 PRINT "ERROR IN ENTERING ELEMENTS":END
710 REM SERIES RESISTOR
720 R1=VR:X1=VX:R2=R:X2=X:GOSUB 1990
730 GOSUB 1860:R=R+1/VAE(I):R2=R:X2=X:GOSUB 1990
740 GOSUB 1860:GOTO 1100
750 REM
760 REM SHUNT RESISTOR
770 R=R+VAE(I):GOTO 1100
780 REM
790 REM UNIT ELEMENT
800 OHM=PI/2*F/EFR(I):C=COS(OHM):S=SIN(OHM):Z=VAE(I):GOSUB
    2680
810 W=TAN(OHM):GOSUB 1920:GOTO 1100
820 REM
830 REM SERIES INDUCTOR
840 R1=VR:X1=VX:R2=R:X2=X:GOSUB 1990
850 GOSUB 1860:X=X+1/VAE(I)*TAN(PI/2*F/EFR(I)):R2=R:X2=X:
    GOSUB 1990
860 GOSUB 1860:GOTO 1100
870 REM
```

```
880 REM SHUNT INDUCTOR
890 X=X-VAE(I)*1/TAN(PI/2*F/EFR(I)):GOTO 1100
900 REM
910 REM SERIES CAPACITOR
920 R1=VR:X1=VX:R2=R:X2=X:GOSUB 1990
930 GOSUB 1860:X=X-1/VAE(I)*1/TAN(PI/2*F/EFR(I)):R2=R:X2=X:
    GOSUB 1990
940 GOSUB 1860:GOTO 1100
950 REM
960 REM SHUNT CAPACITOR
970 X=X+VAE(I)*TAN(PI/2*F/EFR(I)):GOTO 1100
980 REM
990 REM BRANCH DEFINED
1000 RWAIT=R:XWAIT=X:R=0:X=0:GOTO 1100
1010 REM
1020 REM END SERIES BRANCH DEFINED
1030 R1=VR:X1=VX:R2=RWAIT:X2=XWAIT:GOSUB 1990
1040 GOSUB 1860:RHOLD=R:XHOLD=X:R=RWAIT:X=XWAIT:GOSUB 1860
1050 R=R+RHOLD:X=X+XHOLD:R2=R:X2=X:GOSUB 1990
1060 GOSUB 1860:GOTO 1100
1070 REM END PARALLEL BRANCH DEFINED
1080 R=R+RWAIT:X=X+XWAIT:GOTO 1100
1090 REM
1100 NEXT I
1110 REM EVEN MODE ADMITTANCE AND VOLTAGE CALCULATED
1120 RYE(J)=R:XYE(J)=X:VRE(J)=VR:VXE(J)=VX
1130 R=0:X=0:VR=1:VX=0
1140 FOR I=1 TO M
1150 IF TYO$(I)="SR" OR TYO$(I)="sr" THEN GOTO 1210 ELSE IF
     TYO$(I)="PR" OR TYO$(I)="pr" THEN GOTO 1260
1160 IF TYO$(I)="U" OR TYO$(I)="u" THEN GOTO 1290 ELSE IF
     TYO$(I)="SI" OR TYO$(I)="si" THEN GOTO 1330
1170 IF TYO$(I)="PI" OR TYO$(I)="pi" THEN GOTO 1380 ELSE IF
     TYO$(I)="SC" OR TYO$(I)="sc" THEN GOTO 1410
1180 IF TYO$(I)="PC" OR TYO$(I)="pc" THEN GOTO 1460 ELSE IF
     TYO$(I)="B" OR TYO$(I)="b" THEN GOTO 1480
1190 IF TYO$(I)="SB" OR TYO$(I)="sb" THEN GOTO 1510 ELSE IF
     TYO$(I)="PB" OR TYO$(I)="pb" THEN GOTO 1570
1200 PRINT "ERROR IN ENTERING ELEMENTS":END
1210 REM SERIES RESISTOR
1220 R1=VR:X1=VX:R2=R:X2=X:GOSUB 1990
1230 GOSUB 1860:R=R+1/VAO(I):R2=R:X2=X:GOSUB 1990
1240 GOSUB 1860:GOTO 1600
1250 REM
1260 REM SHUNT RESISTOR
1270 R=R+VAO(I):GOTO 1600
1280 REM
1290 REM UNIT ELEMENT
1300 OHM=PI/2*F/OFR(I):C=COS(OHM):S=SIN(OHM):Z=VAO(I):GOSUB
     2680
1310 W=TAN(OHM):GOSUB 1920:GOTO 1600
1320 REM
```

```
1330 REM SERIES INDUCTOR
1340 R1=VR:X1=VX:R2=R:X2=X:GOSUB 1990
1350 GOSUB 1860:X=X+1/VAO(I)*TAN(PI/2*F/OFR(I)):R2=R:X2=X:
     GOSUB 1990
1360 GOSUB 1860:GOTO 1600
1370 REM
1380 REM SHUNT INDUCTOR
1390 X=X-VAO(I)*1/TAN(PI/2*F/OFR(I)):GOTO 1600
1400 REM
1410 REM SERIES CAPACITOR
1420 R1=VR:X1=VX:R2=R:X2=X:GOSUB 1990
1430 GOSUB 1860:X=X-1/VAO(I)*1/TAN(PI/2*F/OFR(I)):R2=R:X2=X:
     GOSUB 1990
1440 GOSUB 1860:GOTO 1600
1450 REM
1460 REM SHUNT CAPACITOR
1470 X=X+VAO(I)*TAN(PI/2*F/OFR(I)):GOTO 1600
1480 REM BRANCH DEFINED
1490 RWAIT=R:XWAIT=X:R=0:X=0:GOTO 1600
1500 REM
1510 REM END SERIES BRANCH DEFINED
1520 R1=VR:X1=VX:R2=RWAIT:X2=XWAIT:GOSUB 1990
1530 GOSUB 1860:RHOLD=R:XHOLD=X:R=RWAIT:X=XWAIT:GOSUB 1860
1540 R=R+RHOLD:X=X+XHOLD:R2=R:X2=X:GOSUB 1990
1550 GOSUB 1860:GOTO 1600
1560 REM
1570 REM END PARALLEL BRANCH DEFINED
1580 R=R+RWAIT:X=X+XWAIT:GOTO 1600
1590 REM
1600 NEXT I
1610 REM ODD MODE ADMITTANCE AND VOLTAGE CALCULATED
1620 RYO(J)=R:XYO(J)=X:VRO(J)=VR:VXO(J)=VX
1630 REM NOW FOLLOWS THE CALCULATION OF GAMMA EVEN & ODD.
1640 R=RYE(J)+1:X=XYE(J):GOSUB 1860:R1=R:X1=X:R2=1-RYE(J):
     X2=-XYE(J)
1650 GOSUB 1990:RGE(J)=R1:XGE(J)=X1
1660 REM RGE(J) AND XGE(J) ARE THE REAL & IMAG PARTS OF GAMMA
     EVEN
1670 R=RYO(J)+1:X=XYO(J):GOSUB 1860:R1=R:X1=X:R2=1-RYO(J):
     X2=-XYO(J)
1680 GOSUB 1990:RGO(J)=R1:XGO(J)=X1
1690 REM RGO(J) AND XGO(J) ARE THE REAL & IMAG PARTS OF GAMMA
     ODD
1700 R=RGE(J)+1:X=XGE(J):GOSUB 1860:R1=R:X1=X:R2=VRE(J):
     X2=VXE(J)
1710 GOSUB 1990:VRE(J)=R1:VXE(J)=X1
1720 R=VRE(J):X=VXE(J):GOSUB 1860:VRE(J)=R:VXE(J)=X
1730 R=RGO(J)+1:X=XGO(J):GOSUB 1860:R1=R:X1=X:R2=VRO(J):
     X2=VXO(J)
1740 GOSUB 1990:VRO(J)=R1:VXO(J)=X1
1750 R=VRO(J):X=VXO(J):GOSUB 1860:VRO(J)=R:VXO(J)=X
1760 REM VRE,VXE,VRO,VXO ARE THE REAL & IMAG PARTS OF E & O
```

```
     TRANS.
1770 REM REAL AND IM PARTS OF GAMMA AND TRANS CALCULATED
1780 NEXT J
1790 BEEP:CLS
1800 REM PLOTS FOLLOW
1810 XINDEEL=10:YINDEEL=5:YBEGIN=-50:YEINDE=0:GOSUB 2300
1820 INPUT "WHICH PARAMETER WOULD YOU LIKE TO SEE";WYS
1830 IF WYS=1 THEN GOTO 2050 ELSE IF WYS=2 THEN GOTO 2090
1840 IF WYS=3 THEN GOTO 2130 ELSE IF WYS=4 THEN GOTO 2170
1850 END
1860 REM ***************************************
1870 REM *                INVERSE              *
1880 REM ***************************************
1890 Q1=R/(R^2+X^2):Q2=-X/(R^2+X^2)
1900 R=Q1:X=Q2
1910 RETURN
1920 REM ***************************************
1930 REM *          TRANSMISSION LINE          *
1940 REM ***************************************
1950 Q3=Z/((Z-W*X)^2+(W*R)^2)
1960 Q4=Q3*(R*(Z-W*X)+W*R*(X+W*Z)):
     Q5=Q3*((Z-W*X)*(X+W*Z)-W*R^2)
1970 R=Q4:X=Q5
1980 RETURN
1990 REM ***************************************
2000 REM *                PRODUCT              *
2010 REM ***************************************
2020 Q6=R1*R2-X1*X2:Q7=R1*X2+R2*X1:R1=Q6:X1=Q7
2030 RETURN
2040 REM START OF CALCULATION OF S-PARAMETERS
2050 FOR J=XBEGIN TO XEINDE STEP XSTAP
2060 SSS=((RGE(J)+RGO(J))^2+(XGE(J)+XGO(J))^2)/4
2070 HOR=J:VERT=4.343*LOG(SSS):GOSUB 2600:NEXT J
2080 GOTO 2210
2090 FOR J=XBEGIN TO XEINDE STEP XSTAP
2100 SSS=((RGE(J)-RGO(J))^2+(XGE(J)-XGO(J))^2)/4
2110 HOR=J:VERT=4.343*LOG(SSS):GOSUB 2600:NEXT J
2120 GOTO 2210
2130 FOR J=XBEGIN TO XEINDE STEP XSTAP
2140 SSS=((VRE(J)+VRO(J))^2+(VXE(J)+VXO(J))^2)/4
2150 HOR=J:VERT=4.343*LOG(SSS):GOSUB 2600:NEXT J
2160 GOTO 2210
2170 FOR J=XBEGIN TO XEINDE STEP XSTAP
2180 SSS=((VRE(J)-VRO(J))^2+(VXE(J)-VXO(J))^2)/4
2190 HOR=J:VERT=4.343*LOG(SSS):GOSUB 2600:NEXT J
2200 GOTO 2210
2210 INPUT "TO REVIEW WITH NEW SCALE, TYPE # OF dB, OTHERWISE
     0";CHNG
2220 IF CHNG=0 THEN GOTO 1800
2230 IF CHNG<20 THEN GOTO 2260
2240 YBEGIN=-CHNG:YINDEEL=-YBEGIN/10:GOSUB 2280
2250 GOTO 1820
```

```
2260 YBEGIN=-CHNG:YINDEEL=-YBEGIN:GOSUB 2280
2270 GOTO 1820
2280 INPUT "CLEAR SCREEN";VEE:IF VEE=0 THEN 2300
2290 CLS
2300 REM ****************************************
2310 REM *                PLOT1                 *
2320 REM ****************************************
2330 REM Subroetine om grafieke te teken.
2340 REM Modelnommer 2.2
2350 PI=3.1415926535#
2360 SCREEN 105:XNUL=25:YNUL=25:XVELD=615:YVELD=300
2370 XWYDTE=XVELD-XNUL:YWYDTE=YVELD-YNUL
2380 REM INDIEN VASTE ROOSTER BENODIG WORD, VOLG LYNE
2390 GOTO 2430
2400 INPUT "AANTAL INDELINGS IN X EN Y";XINDEEL,YINDEEL
2410 INPUT "BEGIN-EN EINDWAARDES VAN X";XBEGIN,XEINDE
2420 INPUT "BEGIN-EN EINDWAARDES VAN Y";YBEGIN,YEINDE
2430 REM
2440 LINE (XVELD,YVELD)-(XNUL,YNUL),1,B
2450 FOR N=1 TO XINDEEL-1
2460 LINE (XNUL+N*XWYDTE/XINDEEL,YVELD)-
     (XNUL+N*XWYDTE/XINDEEL,YNUL),1
2470 NEXT N
2480 FOR N=1 TO YINDEEL-1
2490 LINE (XNUL,YNUL+N*YWYDTE/YINDEEL)-
     (XVELD,YNUL+N*YWYDTE/YINDEEL),1
2500 NEXT N:PRINT "X-RANGE = "XBEGIN" TO "XEINDE
2510 PRINT "Y-RANGE = "YBEGIN" TO "YEINDE:PSET (25,25)
2520 GOTO 2650
2530 REM
2540 REM ****************************************
2550 REM TEKENSUBROETINE VOLG.  GEBRUIK DEUR TE ROEP:
2560 REM   "VERT=..VERTIKALE FUNKSIE"
2570 REM   "HOR=..HORISONTALE FUNKSIE"
2580 REM SIT DAN 'N GOSUB IN VIR LYN 400
2590 REM SUBROETINE VIR SKALEER EN TEKEN
2600 REM ****************************************
2610 HOR=(HOR-XBEGIN)*XWYDTE/(XEINDE-XBEGIN)+25
2620 VERT=YVELD-(VERT-YBEGIN)*YWYDTE/(YEINDE-YBEGIN)
2630 LINE -(HOR,VERT)
2640 RETURN
2650 REM EERSTE LYN VAN DIE PROGRAM IS LYN 500.
2660 REM ****************************************
2670 RETURN
2680 REM ****************************************
2690 REM *               V-LINE                 *
2700 REM ****************************************
2710 QR=VR*C-VR*X*S/Z-VX*R*S/Z
2720 QX=VX*C-VX*X*S/Z+VR*R*S/Z
2730 VR=QR:VX=QX
2740 RETURN
```

A4.2 SYMTHREE

```
10 REM **************************************
20 REM *                                    *
30 REM *              SYMTHREE              *
40 REM *                                    *
50 REM **************************************
60 REM
70 PI=3.1415926535#
80 REM Version # 5.04     Date:  10-04-1988
90 REM Program analyzes cascaded threeports
100 REM by means of odd and even mode networks.
110 CLS
120 PRINT " "
130 PRINT "                              **************"
140 PRINT "Threeport Analysis Program   *  SYMTHREE  *"
150 PRINT "                              **************"
160 PRINT "Element Codes:"
170 PRINT " "
180 PRINT "      N = Order of Network"
190 PRINT "     SR = Series Resistor"
200 PRINT "     PR = Shunt Resistor"
210 PRINT "      U = Unit Element"
220 PRINT "     SI = Series Inductor"
230 PRINT "     PI = Shunt Inductor"
240 PRINT "     SC = Series Capacitor"
250 PRINT "     PC = Shunt Capacitor"
260 PRINT "      B = Start of a Branch Circuit"
270 PRINT "     SB = End of Branch, connected in Series"
280 PRINT "     PB = End of Branch, connected in Parallel"
290 PRINT " "
300 PRINT "********************************************"
310 PRINT " "
320 PRINT "J.A.G. Malherbe, University of Pretoria, 1987"
330 PRINT " "
340 PRINT "********************************************"
350 REM
360 REM START OF DATA INPUT
370 DIM RYE(150),XYE(150),RYO(150),XYO(150),RGE(150),XGE(150)
380 DIM RGO(150),XGO(150),VRE(150),VXE(150),VRO(150),VXO(150)
390 INPUT "ORDER OF NETWORK, M=";M
400 PRINT "*****************************"
410 PRINT "NOW TYPE IN THE ELEMENT TYPES, FOLLOWED BY THEIR "
420 PRINT "ADMITTANCE VALUES, AND RESONANT FREQUENCIES"
430 PRINT "*****************************"
440 PRINT "EVEN MODE NETWORK:
450 PRINT "****************************
460 FOR I=1 TO M:INPUT TYE$(I),VAE(I),EFR(I):NEXT I
```

```
470 PRINT "******************************"
480 PRINT "END OF EVEN MODE NETWORK"
490 PRINT "******************************"
500 REM
510 PRINT "******************************"
520 PRINT "ODD MODE NETWORK:"
530 PRINT "******************************"
540 FOR I=1 TO M:INPUT TYO$(I),VAO(I),OFR(I):NEXT I
550 PRINT "******************************"
560 PRINT "END OF ODD MODE NETWORK"
570 PRINT "******************************"
580 PRINT "CALCULATING ........"
590 XBEGIN=50:XEINDE=150:XSTAP=3
600 FOR J=XBEGIN TO XEINDE STEP XSTAP:F=.01*J+1E-13
610 REM CALCULATION OF YE NOW FOLLOWS
620 REM
630 R=0:X=0:VR=1:VX=0
640 FOR I=1 TO M
650 IF TYE$(I)="SR" OR TYE$(I)="sr" THEN GOTO 710 ELSE IF
    TYE$(I)="PR" OR TYE$(I)="pr" THEN GOTO 760
660 IF TYE$(I)="U" OR TYE$(I)="u" THEN GOTO 790 ELSE IF
    TYE$(I)="SI" OR TYE$(I)="si" THEN GOTO 830
670 IF TYE$(I)="PI" OR TYE$(I)="pi" THEN GOTO 880 ELSE IF
    TYE$(I)="SC" OR TYE$(I)="sc" THEN GOTO 910
680 IF TYE$(I)="PC" OR TYE$(I)="pc" THEN GOTO 960 ELSE IF
    TYE$(I)="B" OR TYE$(I)="b" THEN GOTO 990
690 IF TYE$(I)="SB" OR TYE$(I)="sb" THEN GOTO 1020 ELSE IF
    TYE$(I)="PB" OR TYE$(I)="pb" THEN GOTO 1070
700 PRINT "ERROR IN ENTERING ELEMENTS":END
710 REM SERIES RESISTOR
720 R1=VR:X1=VX:R2=R:X2=X:GOSUB 1990
730 GOSUB 1860:R=R+1/VAE(I):R2=R:X2=X:GOSUB 1990
740 GOSUB 1860:GOTO 1100
750 REM
760 REM SHUNT RESISTOR
770 R=R+VAE(I):GOTO 1100
780 REM
790 REM UNIT ELEMENT
800 OHM=PI/2*F/EFR(I):C=COS(OHM):S=SIN(OHM):Z=VAE(I):GOSUB
    2830
810 W=TAN(OHM):GOSUB 1920:GOTO 1100
820 REM
830 REM SERIES INDUCTOR
840 R1=VR:X1=VX:R2=R:X2=X:GOSUB 1990
850 GOSUB 1860:X=X+1/VAE(I)*TAN(PI/2*F/EFR(I)):R2=R:X2=X:
    GOSUB 1990
860 GOSUB 1860:GOTO 1100
870 REM
880 REM SHUNT INDUCTOR
890 X=X-VAE(I)*1/TAN(PI/2*F/EFR(I)):GOTO 1100
900 REM
910 REM SERIES CAPACITOR
```

```
920 R1=VR:X1=VX:R2=R:X2=X:GOSUB 1990
930 GOSUB 1860:X=X-1/VAE(I)*1/TAN(PI/2*F/EFR(I)):R2=R:X2=X:
    GOSUB 1990
940 GOSUB 1860:GOTO 1100
950 REM
960 REM SHUNT CAPACITOR
970 X=X+VAE(I)*TAN(PI/2*F/EFR(I)):GOTO 1100
980 REM
990 REM BRANCH DEFINED
1000 RWAIT=R:XWAIT=X:R=0:X=0:GOTO 1100
1010 REM
1020 REM END SERIES BRANCH DEFINED
1030 R1=VR:X1=VX:R2=RWAIT:X2=XWAIT:GOSUB 1990
1040 GOSUB 1860:RHOLD=R:XHOLD=X:R=RWAIT:X=XWAIT:GOSUB 1860
1050 R=R+RHOLD:X=X+XHOLD:R2=R:X2=X:GOSUB 1990
1060 GOSUB 1860:GOTO 1100
1070 REM END PARALLEL BRANCH DEFINED
1080 R=R+RWAIT:X=X+XWAIT:GOTO 1100
1090 REM
1100 NEXT I
1110 REM EVEN MODE ADMITTANCE AND VOLTAGE CALCULATED
1120 RYE(J)=R:XYE(J)=X:VRE(J)=VR:VXE(J)=VX
1130 R=0:X=0:VR=1:VX=0
1140 FOR I=1 TO M
1150 IF TYO$(I)="SR" OR TYO$(I)="sr" THEN GOTO 1210 ELSE IF
     TYO$(I)="PR" OR TYO$(I)="pr" THEN GOTO 1260
1160 IF TYO$(I)="U" OR TYO$(I)="u" THEN GOTO 1290 ELSE IF
     TYO$(I)="SI" OR TYO$(I)="si" THEN GOTO 1330
1170 IF TYO$(I)="PI" OR TYO$(I)="pi" THEN GOTO 1380 ELSE IF
     TYO$(I)="SC" OR TYO$(I)="sc" THEN GOTO 1410
1180 IF TYO$(I)="PC" OR TYO$(I)="pc" THEN GOTO 1460 ELSE IF
     TYO$(I)="B" OR TYO$(I)="b" THEN GOTO 1480
1190 IF TYO$(I)="SB" OR TYO$(I)="sb" THEN GOTO 1510 ELSE IF
     TYO$(I)="PB" OR TYO$(I)="pb" THEN GOTO 1570
1200 PRINT "ERROR IN ENTERING ELEMENTS":END
1210 REM SERIES RESISTOR
1220 R1=VR:X1=VX:R2=R:X2=X:GOSUB 1990
1230 GOSUB 1860:R=R+1/VAO(I):R2=R:X2=X:GOSUB 1990
1240 GOSUB 1860:GOTO 1600
1250 REM
1260 REM SHUNT RESISTOR
1270 R=R+VAO(I):GOTO 1600
1280 REM
1290 REM UNIT ELEMENT
1300 OHM=PI/2*F/OFR(I):C=COS(OHM):S=SIN(OHM):Z=VAO(I):
     GOSUB 2830
1310 W=TAN(OHM):GOSUB 1920:GOTO 1600
1320 REM
1330 REM SERIES INDUCTOR
1340 R1=VR:X1=VX:R2=R:X2=X:GOSUB 1990
1350 GOSUB 1860:X=X+1/VAO(I)*TAN(PI/2*F/OFR(I)):R2=R:X2=X:
     GOSUB 1990
```

```
1360 GOSUB 1860:GOTO 1600
1370 REM
1380 REM SHUNT INDUCTOR
1390 X=X-VAO(I)*1/TAN(PI/2*F/OFR(I)):GOTO 1600
1400 REM
1410 REM SERIES CAPACITOR
1420 R1=VR:X1=VX:R2=R:X2=X:GOSUB 1990
1430 GOSUB 1860:X=X-1/VAO(I)*1/TAN(PI/2*F/OFR(I)):R2=R:X2=X:
     GOSUB 1990
1440 GOSUB 1860:GOTO 1600
1450 REM
1460 REM SHUNT CAPACITOR
1470 X=X+VAO(I)*TAN(PI/2*F/OFR(I)):GOTO 1600
1480 REM BRANCH DEFINED
1490 RWAIT=R:XWAIT=X:R=0:X=0:GOTO 1600
1500 REM
1510 REM END SERIES BRANCH DEFINED
1520 R1=VR:X1=VX:R2=RWAIT:X2=XWAIT:GOSUB 1990
1530 GOSUB 1860:RHOLD=R:XHOLD=X:R=RWAIT:X=XWAIT:GOSUB 1860
1540 R=R+RHOLD:X=X+XHOLD:R2=R:X2=X:GOSUB 1990
1550 GOSUB 1860:GOTO 1600
1560 REM
1570 REM END PARALLEL BRANCH DEFINED
1580 R=R+RWAIT:X=X+XWAIT:GOTO 1600
1590 REM
1600 NEXT I
1610 REM ODD MODE ADMITTANCE AND VOLTAGE CALCULATED
1620 RYO(J)=R:XYO(J)=X:VRO(J)=VR:VXO(J)=VX
1630 REM NOW FOLLOWS THE CALCULATION OF GAMMA EVEN & ODD.
1640 R=RYE(J)+1:X=XYE(J):GOSUB 1860:R1=R:X1=X:R2=1-RYE(J):
     X2=-XYE(J)
1650 GOSUB 1990:RGE(J)=R1:XGE(J)=X1
1660 REM RGE(J) AND XGE(J) ARE THE REAL & IMAG PARTS OF GAMMA
     EVEN
1670 R=RYO(J)+1:X=XYO(J):GOSUB 1860:R1=R:X1=X:R2=1-RYO(J):
     X2=-XYO(J)
1680 GOSUB 1990:RGO(J)=R1:XGO(J)=X1
1690 REM RGO(J) AND XGO(J) ARE THE REAL & IMAG PARTS OF GAMMA
     ODD
1700 R=RGE(J)+1:X=XGE(J):GOSUB 1860:R1=R:X1=X:R2=VRE(J):
     X2=VXE(J)
1710 GOSUB 1990:VRE(J)=R1:VXE(J)=X1
1720 R=VRE(J):X=VXE(J):GOSUB 1860:VRE(J)=R:VXE(J)=X
1730 R=RGO(J)+1:X=XGO(J):GOSUB 1860:R1=R:X1=X:R2=VRO(J):
     X2=VXO(J)
1740 GOSUB 1990:VRO(J)=R1:VXO(J)=X1
1750 R=VRO(J):X=VXO(J):GOSUB 1860:VRO(J)=R:VXO(J)=X
1760 REM VRE,VXE,VRO,VXO ARE THE REAL & IMAG PARTS OF E & O
     TRANS.
1770 REM REAL AND IM PARTS OF GAMMA AND TRANS CALCULATED
1780 NEXT J
1790 BEEP:CLS
```

```
1800 REM PLOTS FOLLOW
1810 XINDEEL=10:YINDEEL=5:YBEGIN=-50:YEINDE=0:GOSUB 2300
1820 INPUT "WHICH PARAMETER WOULD YOU LIKE TO SEE";WYS
1830 IF WYS=1 THEN GOTO 2050 ELSE IF WYS=2 THEN GOTO 2090
1840 IF WYS=3 THEN GOTO 2130 ELSE IF WYS=4 THEN GOTO 2170
1850 END
1860 REM *************************************
1870 REM *                  INVERSE                  *
1880 REM *************************************
1890 Q1=R/(R^2+X^2):Q2=-X/(R^2+X^2)
1900 R=Q1:X=Q2
1910 RETURN
1920 REM *************************************
1930 REM *            TRANSMISSION LINE            *
1940 REM *************************************
1950 Q3=Z/((Z-W*X)^2+(W*R)^2)
1960 Q4=Q3*(R*(Z-W*X)+W*R*(X+W*Z)):
     Q5=Q3*((Z-W*X)*(X+W*Z)-W*R^2)
1970 R=Q4:X=Q5
1980 RETURN
1990 REM *************************************
2000 REM *                  PRODUCT                  *
2010 REM *************************************
2020 Q6=R1*R2-X1*X2:Q7=R1*X2+R2*X1:R1=Q6:X1=Q7
2030 RETURN
2040 REM START OF CALCULATION OF S-PARAMETERS
2050 FOR J=XBEGIN TO XEINDE STEP XSTAP
2060 SSS=(RGE(J)^2+XGE(J)^2)
2070 HOR=J:VERT=4.343*LOG(SSS):GOSUB 2650:NEXT J
2080 GOTO 2210
2090 FOR J=XBEGIN TO XEINDE STEP XSTAP
2100 SSS=(1-RGE(J)^2-XGE(J)^2)/2
2110 HOR=J:VERT=4.343*LOG(SSS):GOSUB 2650:NEXT J
2120 GOTO 2210
2130 FOR J=XBEGIN TO XEINDE STEP XSTAP
2140 SSS=((RGE(J)+RGO(J))^2+(XGE(J)+XGO(J))^2)/4
2150 HOR=J:VERT=4.343*LOG(SSS):GOSUB 2650:NEXT J
2160 GOTO 2210
2170 FOR J=XBEGIN TO XEINDE STEP XSTAP
2180 SSS=((RGE(J)-RGO(J))^2+(XGE(J)-XGO(J))^2)/4
2190 HOR=J:VERT=4.343*LOG(SSS):GOSUB 2650:NEXT J
2200 GOTO 2210
2210 INPUT "TO REVIEW WITH NEW SCALE, TYPE # OF dB,
     OTHERWISE 0";CHNG
2220 IF CHNG=0 THEN GOTO 1800
2230 IF CHNG<20 THEN GOTO 2260
2240 YBEGIN=-CHNG:YINDEEL=-YBEGIN/10:GOSUB 2280
2250 GOTO 1820
2260 YBEGIN=-CHNG:YINDEEL=-YBEGIN:GOSUB 2280
2270 GOTO 1820
2280 INPUT "CLEAR SCREEN";VEE:IF VEE=0 THEN 2300
2290 CLS
2300 REM *************************************
2310 REM *                   PLOT1                   *
2320 REM *************************************
2330 REM Subroetine om grafieke te teken.
```

```
2340 REM Modelnommer 2.2
2350 PI=3.1415926535#
2360 SCREEN 105:XNUL=25:YNUL=25:XVELD=615:YVELD=300
2370 XWYDTE=XVELD-XNUL:YWYDTE=YVELD-YNUL
2380 REM INDIEN VASTE ROOSTER BENODIG WORD, VOLG LYNE
2390 GOTO 2430
2400 INPUT "AANTAL INDELINGS IN X EN Y";XINDEEL,YINDEEL
2410 INPUT "BEGIN-EN EINDWAARDES VAN X";XBEGIN,XEINDE
2420 INPUT "BEGIN-EN EINDWAARDES VAN Y";YBEGIN,YEINDE
2430 REM
2440 LINE (XVELD,YVELD)-(XNUL,YNUL),1,B
2450 FOR N=1 TO XINDEEL-1
2460 LINE (XNUL+N*XWYDTE/XINDEEL,YVELD)-
     (XNUL+N*XWYDTE/XINDEEL,YNUL),1
2470 NEXT N
2480 FOR N=1 TO YINDEEL-1
2490 LINE (XNUL,YNUL+N*YWYDTE/YINDEEL)-
     (XVELD,YNUL+N*YWYDTE/YINDEEL),1
2500 NEXT N:PRINT "X-RANGE = "XBEGIN" TO "XEINDE
2510 PRINT "Y-RANGE = "YBEGIN" TO "YEINDE:PSET (25,25)
2520 GOTO 2800
2530 REM
2540 REM ****************************************
2550 REM TEKENSUBROETINE VOLG.  GEBRUIK DEUR TE ROEP:
2560 REM   "VERT=..VERTIKALE FUNKSIE"
2570 REM   "HOR=..HORISONTALE FUNKSIE"
2580 REM SIT DAN 'N GOSUB IN VIR LYN 400
2590 REM SUBROETINE VIR SKALEER EN TEKEN
2600 REM ****************************************
2610 HOR=(HOR-XBEGIN)*XWYDTE/(XEINDE-XBEGIN)+25
2620 VERT=YVELD-(VERT-YBEGIN)*YWYDTE/(YEINDE-YBEGIN)
2630 IF VERT>YVELD THEN 2730
2640 IF VERTH>YVELD THEN 2760
2650 HORH=HOR:VERTH=VERT
2660 LINE -(HOR,VERT)
2670 GOTO 2790
2680 VERTN=YVELD:HORN=HORH+(HOR-HORH)*
     ABS((YVELD-VERTH)/(VERT-VERTH))
2690 HORH=HOR:VERTH=VERT
2700 LINE -(HORN,VERTN):GOTO 2790
2710 VERTN=YVELD:HORN=HORH+(HOR-HORH)*
     ABS((YVELD-VERTH)/(VERT-VERTH))
2720 HORH=HOR:VERTH=VERT
2730 LINE (HORN,VERTN)-(HOR,VERT)
2740 RETURN
2750 REM EERSTE LYN VAN DIE PROGRAM IS LYN 500.
2760 REM ****************************************
2770 RETURN
2780 REM ***************************************
2790 REM *                V-LINE               *
2800 REM ***************************************
2810 QR=VR*C-VR*X*S/Z-VX*R*S/Z
2820 QX=VX*C-VX*X*S/Z+VR*R*S/Z
2830 VR=QR:VX=QX
2840 RETURN
```

Appendix B
Coupled-Line Parameters

B1 INTRODUCTION

Basic to the design and realization of a very large number of microwave circuits is the realization of coupled-line sections. This appendix discusses a variety of methods for the realization of coupled lines. As the sole intention is the application of such information to network design, we study no derivations or give any proofs; if necessary, the interested reader can consult the references.

In the section following this introduction, a number of typical and popular materials commonly employed in microwave component manufacture are considered briefly and the limits on chemical machinability of lines and gaps is discussed. Other sections treat specific physical cross sections individually, list their design equations, and discuss realizability.

As we said in the preface, the aim of this appendix is to provide a ready access to first-order design equations. Although, in many cases, they will be found to be as accurate, the aim is to provide some feel for the parameters rather than a computer-aided design package. In the same spirit, we provide a very brief section on the calculation of single, uncoupled lines at the end of the material on coupled lines.

The design equations have been programmed into a simple computer design program that makes it possible to skip the step from network properties to physical dimensions, and this must be seen in the same light as the rest of the material in the appendix. The listing of the program is provided at the end of this appendix.

B2 MATERIALS

A very large range of materials is available to the microwave engineer, and it would be outside of the scope of this book to attempt to list and evaluate them. Very briefly, and as an example of the arguments, we will discuss a few of the parameters.

B2.1 Conductor Thickness

Typical conductor claddings for stripline or microstrip (soft substrates) vary from as low as $\frac{1}{4}$ and $\frac{1}{2}$ oz. to 1 or 2 oz. of copper per square yard. In terms of thickness, 1 oz. of copper corresponds to a conductor thickness of 35 μm. In the photochemical etching process commonly used to manufacture the circuit, a certain amount of undercut always occurs [1], and consequently, the desired linewidths or gaps are not realized.

In practice it turns out that if the narrowest linewidth is limited to about 10 times the copper thickness, the roughness of the etch is negligible. This same criterion also holds for gaps between the lines. It must be stressed that these numbers are only guidelines and are realistic for manufacture with modest control over the etching process or in cases where mass production is being done. With care, the parameters could readily be improved on, and gaps or linewidths of half those just given could be realized.

These restrictions apply equally well to the etching of single lines, thereby setting a maximum impedance level on a single, uncoupled line. In the case of microstrip on an alumina substrate, Muraguchi *et al.* ([4], cited in Chapter 3) state that the useful range of impedances lies between 10 Ω and 100 Ω to between 40 Ω and 160 Ω depending on the substrate thickness.

The maximum linewidth in all cases is determined by the condition for the propagation of the waveguide mode. In the case of stripline, this corresponds to characteristic impedances in the twenties of Ω, but in most cases this is not a crucial factor. In most cases it means that the linewidth must be well below $\lambda/4$.

B2.2 Substrates

The soft dielectric substrates are available in an almost continuous variety from 0.005″ to 0.25″ and more, in steps of 0.005″ or, in metric units, 0.125 mm to 6.35 mm. The dielectric constants vary according to material; the popular substrates are teflon with glass fiber filling, and the dielectric constants vary betweeen $\epsilon_r = 2.17$ through 2.33, 2.45, 2.55, *et cetera*. In most cases the higher values of ϵ_r indicate greater losses as well as a cheaper substrate. Soft substrates for microstrip application include dielectric constants of 6 and 10.25, while alumina has $\epsilon_r = 9.8$.

B3 PARALLEL COUPLED LINES

B3.1 Equations

Stripline, as a microwave printed circuit or in the form of mechanically or chemically machined plates or bars, has always been a popular medium. The geometry is defined in Figure B1. The fringing capacitances for the case of zero thickness strips are given by [2]:

$$\frac{C'_{fe}}{\epsilon}(0, s/b) = \frac{2}{\pi}\ln\left(1 + \tanh\frac{\pi s}{2b}\right) \tag{B.1}$$

$$= \frac{s}{b} - \frac{2}{\pi}\ln\left(\cosh\frac{\pi s}{2b}\right) \tag{B.2}$$

$$\frac{C'_{fo}}{\epsilon}(0, s/b) = \frac{2}{\pi}\ln\left(1 + \coth\frac{\pi s}{2b}\right) \tag{B.3}$$

$$= \frac{s}{b} - \frac{2}{\pi}\ln\left(\sinh\frac{\pi s}{2b}\right) \tag{B.4}$$

where

$$\frac{C'_f(0)}{\epsilon} = \frac{2}{\pi}\ln 2 = 0.4407 \tag{B.5}$$

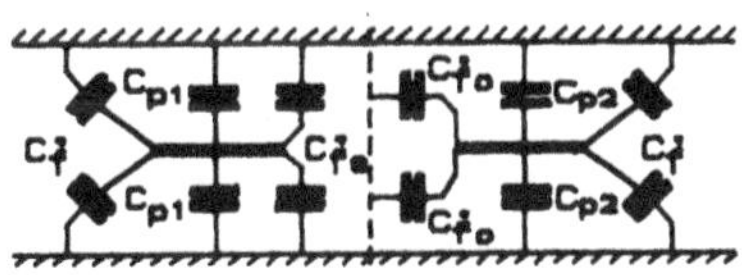

Figure B1 Cross section of a pair of coupled striplines. The plate, fringing, and even- and odd-mode capacitances are shown for the even and odd modes to the left and right of the center line.

For strips of finite thickness, the fringing capacitances are modified:

$$C'_{fe}(t/b,\ s/b) = C'_{fe}(\mathbf{0},\ s/b) \cdot \frac{C'_f(t/b)}{C'_f(\mathbf{0})} \tag{B.6}$$

$$C'_{fo}(t/b,\ s/b) = C'_{fo}(\mathbf{0},\ s/b) \cdot \frac{C'_f(t/b)}{C'_f(\mathbf{0})} \tag{B.7}$$

where

$$C'_f(t/b)/\epsilon = \frac{1}{\pi}\left[2\xi \ln(\xi + 1) - (\xi - 1)\ln(\xi^2 - 1)\right] \tag{B.8}$$

and

$$\xi = \frac{1}{1 - t/b} \tag{B.9}$$

In cases where the strip is not symmetrical to the ground planes, the upper and lower halves are treated separately, and (B.9) is modified to include the difference in distance from the plates. For instance, on the side closer to the ground plane by $d/2$, b is replaced by $(b - d)$. The plate capacitance for each side of the strip is given by

$$C_p/\epsilon = \frac{2w/b}{1 - t/b} \tag{B.10}$$

Once again, b is replaced by the relevant plate distance if the center conductor is not symmetrically placed. The even- and odd-mode impedances are determined from the corresponding capacitances:

$$Z_{0e} = \frac{\eta}{\sqrt{\epsilon_r}} \cdot \frac{1}{C_{0e}/\epsilon} \tag{B.11}$$

$$Z_{0o} = \frac{\eta}{\sqrt{\epsilon_r}} \cdot \frac{1}{C_{0o}/\epsilon} \tag{B.12}$$

where

$$C_{0e}/\epsilon = 2\left(\frac{2w}{b - t} + C'_f/\epsilon + C'_{fe}/\epsilon\right) \tag{B.13}$$

$$C_{0o}/\epsilon = 2\left(\frac{2w}{b-t} + C_f'/\epsilon + C_{fo}'/\epsilon\right) \tag{B.14}$$

The fringing capacitances in (B.13) and (B.14) are of course the relevant values obtained from eqs. (B.1)–(B.5) or (B.6)–(B.9), depending on whether the zero thickness approximation can be used.

The mutual capacitance is obtained from

$$\Delta C/\epsilon = C_{12}/\epsilon = C_{fo}' - C_{fe}' \tag{B.15}$$

$$= \tfrac{1}{2}\left(C_{0o}/\epsilon - C_{0e}/\epsilon\right) \tag{B.16}$$

When three or more center conductors are coupled together, as shown in Figure B2, the two outside conductors have capacitances that can be described by (B.13) and (B.14). The inner conductor, however, will have odd- and even-mode capacitances given by

$$C_{0e2}/\epsilon = 2\left(\frac{2w_2}{b-t} + C_{fe12}'/\epsilon + C_{fe23}'/\epsilon\right) \tag{B.17}$$

$$C_{0o2}/\epsilon = 2\left(\frac{2w_2}{b-t} + C_{fo12}'/\epsilon + C_{fo23}'/\epsilon\right) \tag{B.18}$$

As a rule, the lines in directional couplers are spaced equally (except when the characteristic impedances of the ports are not all equal), so that the even- and odd-mode fringing capacitances between lines 1 and 2, and lines 2 and 3 are the same. In all cases, the characteristic impedances are given by (B.11) and (B.12).

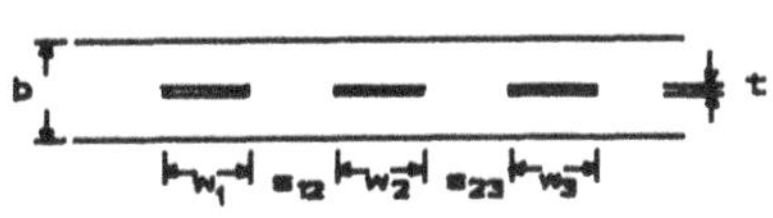

Figure B2 Cross section of coupled striplines with three center conductors.

B3.2 Design Data

The equations in Section B3.1 have been programmed into the computer program LDIMS, a listing of which is included at the end of this appendix together with a simple set of operating instructions. As an example, the procedure to obtain dimensions for a given set of even- an odd-mode impedances are given.

Example

For the design example, we assume that the dimensions for the following set of parameters are needed:

Characteristic impedance,	$Z_0 = 50\ \Omega$
Even-mode impedance,	$Z_{0e} = 60\ \Omega$
Odd-mode impedance,	$Z_{0o} = 41.67\ \Omega$
Ground-plane spacing,	$b = 3.175$ mm
Copper thickness,	$t = 35\ \mu$m
Dielectric constant,	$\epsilon_r = 2.32$

(note that $Z_{0e}Z_{0o} = Z_0^2$)

The computer screen display for this problem is shown in Figure B3. The questions prompted by the program are self-explanatory.

```
RUN
DO YOU WISH TO DESIGN
        (1) A Coupled Coplanar Stripline
        (2) An Offset Parallel-Coupled Pair
        (3) A Coupled Microstripline
        (4) A Symmetrical Stripline
        (5) An Unsymmetrical Stripline
        (6) A Microstripline

Type the number desired, followed by RETURN1

*****************************************************
           COUPLED COPLANAR STRIPLINE
*****************************************************
              Give the Dielectric constant Er=2.32
                the plate separation in mm, b=3.175
                    and conductor thickness t=.035

These parameters are retained.

          Input the even mode impedance, Zoe=60
             and the odd mode impedance, Zoo=41.67

      DIMENSIONS:

                     Spacing, s = .5432227
                   Linewidth, w = 2.272607

          Input the even mode impedance, Zoe=
```

Figure B3 Computer screen display for the calculation of the dimensions of a stripline example.

B3.3 Limits on Coupling

The most serious limitation on the use of stripline in directional coupler realization is the relatively low level of coupling that can be realized. For 3.175 mm ($\frac{1}{8}''$) ground-plane spacing (GPS) and a copper thickness of 35 μm, the realistic realizable gap between two lines is 350 μm. Substitution of $s/b = 0.1102$ into (B.1) and (B.3) gives a maximum value of $\Delta C =$ 1.13. If the gap size is halved, the maximum mutual capacitance is only as much as $\Delta C = 1.56$.

B4 OFFSET PARALLEL-COUPLED STRIPLINES

B4.1 Equations

The offset parallel-coupled striplines allow the coupling to be greatly increased because the two strips do not lie in the same plane, as shown in Figure B4. Very thin layers of dielectric, doubly clad with copper, are commercially available, and the coupled-line pattern is etched on both sides of the center conductor sandwich.

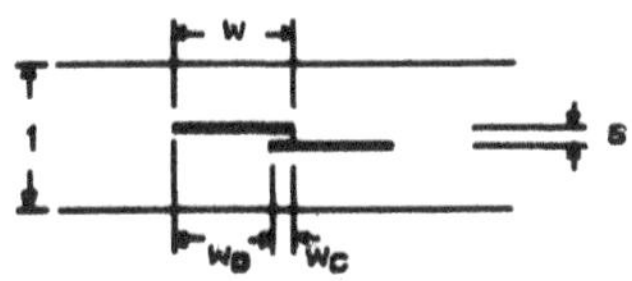

Figure B4 Geometry of offset parallel-coupled stripline.

Shelton [3] gives equations for the capacitances of the line for two cases: when the lines are tightly coupled, and when they are coupled loosely. The former case is depicted in Figure B5(a) and the latter in Figure B5(b). The impedance relations are given by

$$\rho = Z_{0e}/Z_{0o} \tag{B.19}$$

$$Z_{0e} = Z_0\sqrt{\rho} \tag{B.20}$$

$$Z_{0o} = Z_0/\sqrt{\rho} \tag{B.21}$$

and the capacitances are given by the expressions in Sections B4.1.1 and B4.1.2.

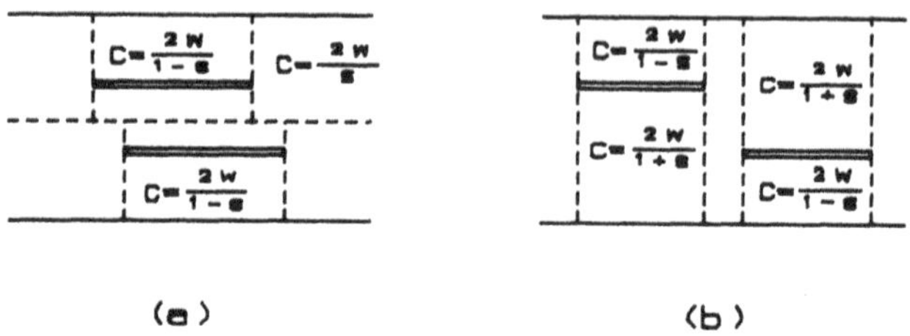

Figure B5 Coupling configuration for (a) tight coupling and (b) loose coupling.

B4.1.1 Tight Coupling

$$C_0 = 2w\left(\frac{1}{1 - s} + \frac{1}{s}\right) + C_{fo}$$

$$= \frac{2w}{s(1 - s)} + C_{fo} \qquad \text{(B.22)}$$

$$C_e = \frac{2w}{1 - s} + C_{fe} \qquad \text{(B.23)}$$

The expressions for C_{fo} and C_{fe} follow, with the collected equations for the tight coupling case.

$$A = \exp\left[\frac{60\pi^2}{\sqrt{\epsilon_r}\, Z_0}\left(\frac{1 - \rho s}{\sqrt{\rho}}\right)\right] \qquad \text{(B.24)}$$

$$B = \frac{pr}{s} = \tfrac{1}{2}(A - 2 + \sqrt{A^2 - 4A}) \qquad \text{(B.25)}$$

$$p = \tfrac{1}{2}\{(B - 1)\tfrac{1}{2}(1 + s) + \sqrt{(B - 1)^2[\tfrac{1}{2}(1 + s)]^2 + 4sB}\} \qquad \text{(B.26)}$$

$$r = \frac{sB}{p} \quad \text{or} \quad r = \frac{p + \tfrac{1}{2}(1 + s)}{1 + \tfrac{1}{2}p(1 + s)/s} \qquad \text{(B.27)}$$

$$C_{fo} = \frac{1}{\pi}\left\{-\frac{2}{1 - s}\log s + \log\left[\frac{pr}{(p + s)(1 + p)(r - s)(1 - r)}\right]\right\} \qquad \text{(B.28)}$$

$$C_{fe} = \frac{1}{\pi}\left\{-\frac{2s}{1-s}\log s - 2\log s + 4\log(s+pr)\right.$$
$$\left. - \log[pr(p+s)(p+1)(r-s)(1-r)]\right\} \tag{B.29}$$

$$C_o = \frac{120\pi\sqrt{\rho}}{Z_o\sqrt{\epsilon_r}} \tag{B.30}$$

$$w = \frac{s(1-s)}{2}(C_0 - C_{fo}) \tag{B.31}$$

$$w_0 = \frac{1}{2\pi}\left\{(1+s)\log\frac{p}{r} + (1-s)\log\left[\frac{(1+p)(r-s)}{(s+p)(1-r)}\right]\right\} \tag{B.32}$$

The condition of the tightest coupling, $w_0 = 0$, is given by

$$\frac{1-\rho_{\max}s}{\sqrt{\rho_{\max}}} = \frac{Z_0\sqrt{\epsilon_r}}{60\pi^2}\log 4 \tag{B.33}$$

B4.1.2 Loose Coupling

The capacitances for loose coupling are given by

$$C_o = 2w\left(\frac{1}{1-s} + \frac{1}{1+s}\right) + C_{fo} - C_f(a=\infty)$$
$$= \frac{4w}{1-s^2} + C_{fo} - C_f(a=\infty) \tag{B.34}$$
$$= \frac{4w}{1-s^2} + C_{fe} - C_f(a=\infty) \tag{B.35}$$

where C_{fe} and C_{fo} are given later in the collected equations for loose coupling, and the fringing capacitance at the end of a single uncoupled strip is $C_f(a = \infty)$.

$$C_0 = \frac{120\pi\sqrt{\rho}}{Z_0\sqrt{\epsilon_r}} \tag{B.36}$$

$$\Delta C = \frac{120\pi}{Z_0\sqrt{\epsilon_r}} \cdot \frac{\rho - 1}{\sqrt{\rho}} \tag{B.37}$$

$$k = \frac{1}{\exp\frac{1}{2}\pi\Delta C - 1} \tag{B.38}$$

$$a = +\sqrt{\frac{(s - k)^2}{(s + 1)^2} + k} - \frac{(s - k)}{(s + 1)} \tag{B.39}$$

$$q = \frac{k}{a} \quad \text{or} \quad q = \tfrac{1}{2}(s + 1)\left[\frac{a + 2s/(s + 1)}{a + (s + 1)/2}\right] \tag{B.40}$$

$$C_{fo} = \frac{2}{\pi}\left[\frac{1}{1 + s}\log\frac{1 + a}{a(1 - q)} - \frac{1}{1 - s}\log q\right] \tag{B.41}$$

$$C_f(a = \infty) = -\frac{2/\pi}{1 + s}\log\tfrac{1}{2}(1 - s) - \frac{2/\pi}{1 - s}\log\tfrac{1}{2}(1 + s) \tag{B.42}$$

$$w_c = \frac{1}{\pi}\left[s\log\frac{q}{a} + (1 - s)\log\left(\frac{1 - q}{1 + a}\right)\right] \tag{B.43}$$

$$w = \frac{1 - s^2}{4}\left[C_o - C_{fo} - C_f(a = \infty)\right] \tag{B.44}$$

In the cases of both the loose and the tight coupling, Shelton [3] states that the following three restrictions apply to the derivation of the equations. In general,

$$\frac{w}{1 - s} \geq 0.35 \tag{B.45}$$

For loose coupling,

$$\frac{2w_0}{1 + s} \geq 0.85 \tag{B.46}$$

For tight coupling,

$$\frac{w_c}{s} \geq 0.7 \tag{B.47}$$

B4.2 Design Data

Graphs of coupling *versus* w/b and w_c/b for various impedance values and different substrates with s/b as a parameter are available in the literature, and Figure B6 shows the typical parametric dependence of the variables [3]. For the purposes of calculating dimensions from known impedance ratios and material properties, the design equations have been programmed into LDIMS. The following example illustrates the use of the program.

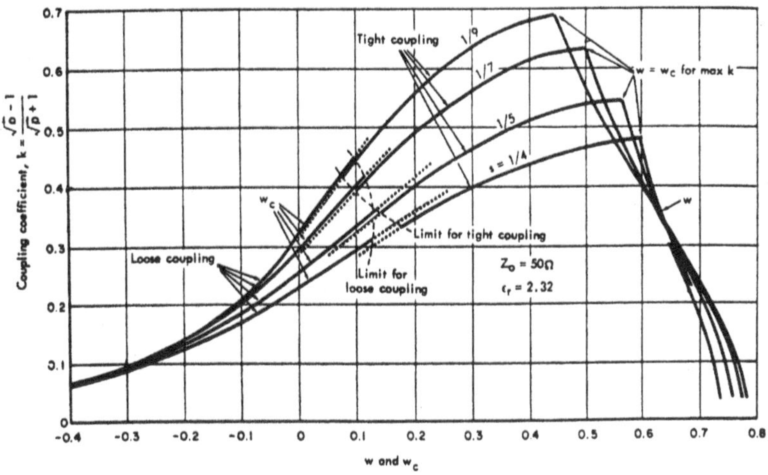

Figure B6 **Graph of coupling *versus* w/b and w_c/b for various impedance values. (© 1966 IEEE, with permission, from [3]).**

Example

For the design example, the following parameters are chosen:

Characteristic impedance,	Z_0	$= 50\ \Omega$
Even-mode impedance,	Z_{0e}	$= 111.80\ \Omega$
Odd-mode impedance,	Z_{0o}	$= 22.36\ \Omega$
Ground-plane spacing,	b	$= 3.300$ mm
Center strip thickness,	t	$= 125\ \mu$m
Center strip ratio,	s	$= 0.03788$
Dielectric constant,	ϵ_r	$= 2.32$

The information was programmed into LDIMS and the relevant computer screen display is shown in Figure B7.

```
RUN
DO YOU WISH TO DESIGN
        (1) A Coupled Coplanar Stripline
        (2) An Offset Parallel-Coupled Pair
        (3) A Coupled Microstripline
        (4) A Symmetrical Stripline
        (5) An Unsymmetrical Stripline
        (6) A Microstripline

Type the number desired, followed by RETURN2

*************************************************
        OFFSET PARALLEL-COUPLED STRIPLINE
*************************************************
               Give the Dielectric constant Er=2.32
                 the plate separation in mm, b=3.3
                 and the center strip ratio, s=.03788

These parameters are retained.

            Input the even mode impedance, Zoe=111.80
               and the odd mode impedance, Zoo=22.36

      DIMENSIONS:

             Coupled width, Wc =          .2789697
                 Line width, W =          1.220193
(Tight coupling)

            Input the even mode impedance, Zoe=
```

Figure B7 Computer screen display for calculating the dimensions of an offset parallel-coupled stripline example.

B4.3 Limits on Coupling

The maximum realizable coupling for the offset line occurs when the lines are coupled directly over each other (i.e., when $w_0 = 0$). For a given characteristic impedance and dielectric constant, the maximum coupling is a function of the center strip thickness. Table B1 shows the relationship between s and ρ_{max} for $Z_\Omega = 50\ \Omega$ and $\epsilon_r = 2.32$ for a number of cases. The last case, that of a ratio of 26.4 corresponds to the case of two thicknesses of 1.59 mm each ($\frac{1}{16}''$) with a layer of 0.125 mm in between (.005″), giving $s = 0.03788$ as in the earlier example.

Table B1 Values of Maximum Coupling for Various Center Strip Ratios

l/s	ρ	ΔC	k	k (dB)
5	3.365	3.190	0.542	−3.52
10	5.731	4.892	0.703	−3.06
15	7.619	5.953	0.768	−2.29
20	9.190	6.687	0.804	−1.90
25	10.533	7.271	0.827	−1.65
26.4	10.877	7.413	0.831	−1.60

B5 PARALLEL COUPLED BARS

If the conductors are allowed to become very thick, as shown in Figure B8, the stripline equations of the previous paragraphs are no longer valid. Under these circumstances, side coupling can be made high and, due to the structural rigidity of the bars, an air dielectric can be used. This in turn allows a substantial increase in the power handling.

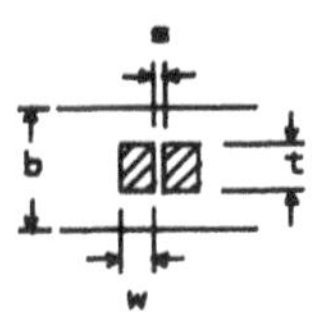

Figure B8 Cross section of thick rectangular bars between parallel plates.

B5.1 Equations

The design approach to utilizing thick coupled lines, is exactly the same as that of the etched stripline except for the determination of the fringing capacitances. The classic work by Getsinger [4] provides curves for obtaining values of the fringing capacitances, if the dimensions are known, or vice versa. However, Perlow [5] presented closed-form equations that are very simple and easy to program on a computer. The equations by

Perlow are, in fact, based on modifications to the thin-strip solutions described earlier. Referring to Figure B8, the mutual capacitance for the thick strip case is

$$\Delta C/\epsilon = \Delta C(t = 0)/\epsilon \cdot A \tag{B.48}$$

$$\Delta C(t = 0)/\epsilon = \frac{2}{\pi} \ln \left[\coth \left(\frac{\pi s}{2b} \right) \right] \tag{B.49}$$

and the strip correction factor is

$$A = \begin{cases} 1.0 + 2.13507 \dfrac{t}{b} \left(\dfrac{s}{b} \right)^{-0.57518} & \dfrac{s}{b} < 0.3 \\ 1.0 + 3.89531 \dfrac{t}{b} \left(\dfrac{s}{b} \right)^{-0.11467} & \dfrac{s}{b} \geq 0.3 \end{cases} \tag{B.50}$$

The fringing capacitance for a semiinfinite strip is the same as for a thin strip, given by (B.8) and (B.9). The even-mode fringing capacitance is given by

$$C_{fe}/\epsilon = A \left\{ \frac{s}{b} - \frac{2}{\pi} \ln \left[\cosh \left(\frac{\pi s}{2b} \right) \right] \right\} + B \tag{B.51}$$

where A is given by (B.50) and B is defined as follows. For

$$t/b \leq 0.5636,$$

$$B = \begin{cases} -\dfrac{t}{b} \left(0.2933 + 3.333 \dfrac{s}{b} \right) & \dfrac{s}{b} < 0.08 \\ -0.56 \dfrac{t}{b} & \dfrac{s}{b} \geq 0.08 \end{cases} \tag{B.52}$$

For

$$t/b > 0.5636$$

$$B = \begin{cases} -0.1653 - 5.6814 \dfrac{s}{b} + 6.7475 \dfrac{s}{b} \left(\dfrac{t}{b} \right) & \dfrac{s}{b} < 0.08 \\ -0.62 + 0.54 \dfrac{t}{b} & \dfrac{s}{b} \geq 0.08 \end{cases} \tag{B.53}$$

The strip width is also compensated for loss of fringe-field capacitance, and the effective width is given by

$$w_{\text{eff}}/b = \begin{cases} \dfrac{w}{b} - w_{c0}\left[1 - \sqrt{\dfrac{w/b \cdot t/b}{0.015}}\right]^2 & \dfrac{w}{b}\left(\dfrac{t}{b}\right) < 0.015 \\ \dfrac{w}{b} & \dfrac{w}{b}\left(\dfrac{t}{b}\right) \geq 0.015 \end{cases} \tag{B.54}$$

where

$$w_{c0} = \frac{2}{\pi}\ln 2 + \frac{w}{b} - \frac{1}{\dfrac{2}{\pi}\ln\left[2\coth\left(\dfrac{\pi w}{4b}\right)\right]} \tag{B.55}$$

Equation (B.55) implies that the correction term is negligibly small if

$$\frac{w}{b} > 0.4 \qquad \text{or} \qquad \frac{w}{b} > \frac{0.015}{t/b}$$

The parallel plate capacitance is related to the effective width by

$$C_p/\epsilon = 2w_{\text{eff}}/(b - t) \tag{B.56}$$

B5.2 Design Data

The equations in B5.1 have been incorporated in the program LDIMS and an example of the calculation follows.

Example

Characteristic impedance,	$Z_0 = 50\ \Omega$
Even-mode impedance,	$Z_{0e} = 111.80\ \Omega$
Odd-mode impedance,	$Z_{0o} = 22.36\ \Omega$
Ground-plane spacing,	$b = 10$ mm
Center strip thickness,	$t = 2$ mm
Dielectric constant,	$\epsilon_r = 1$ (air)

The relevant display for this problem is shown in Figure B9. Information on the odd-mode fringing capacitance, C'_{fo}/ϵ is not used directly in the realization of coupled bars. It is, however, necessary to bring into consideration the effects of the sidewalls that are needed to close in the

```
RUN
DO YOU WISH TO DESIGN
        (1) A Coupled Coplanar Stripline
        (2) An Offset Parallel-Coupled Pair
        (3) A Coupled Microstripline
        (4) A Symmetrical Stripline
        (5) An Unsymmetrical Stripline
        (6) A Microstripline

Type the number desired, followed by RETURN1

**************************************************
           COUPLED COPLANAR STRIPLINE
**************************************************
              Give the Dielectric constant Er=1
                the plate separation in mm, b=10
                    and conductor thickness t=2

These parameters are retained.

          Input the even mode impedance, Zoe=111.8
             and the odd mode impedance, Zoo=22.36

      DIMENSIONS:

                        Spacing, s = .3828125
                      Linewidth, w = 3.753693

          Input the even mode impedance, Zoe=
```

Figure B9 Computer screen display for the calculation of the dimensions of a coupled parallel bar example.

structure. With the thick center conductors used in this type of construction, the coupling to the side walls is substantial, and the fringing capacitance of the end bars is not accurately described by the fringing capacitance C'_e/ϵ. In the place of the latter capacitance, for a gap spacing to the side walls of s_e, the value $C'_{fo}/\epsilon(s_e/2)$ should be used. This can be obtained by subtracting C'_{fe}/ϵ from ΔC.

B5.3 Limits on Coupling

Coupled bars between parallel plates make it possible to realize extremely high values of coupling, as is obvious from the example in the previous paragraph. Specific choices of plate separation and bar thickness will influence the coupling obtainable and will clearly be determined by circumstances.

B6 COUPLED MICROSTRIP LINES

The problems associated with coupled microstrips stem mainly from two sources. First, it is only a pseudo-TEM structure, having a nonhomogeneous cross section. One part of the line has air as a dielectric, while the rest has some type of carrier substrate; popular materials include soft plastics with dielectric constants in the region of 2 to 2.5, composite materials such as mixtures of ceramics and plastics with dielectric constants of as high as 10, and hard substrates such as alumina with a dielectric constant of 9.8. Microstrip lines of course also find application on microwave integrated circuits, using materials such as gallium arsenide.

The nonhomogeneous cross section in Figure B10 shows that the static capacitance approaches for use in the solution of (for instance) stripline cross sectional problems are not as accurate as would be desired. If the frequency is low enough and if the coupler performance demands are not too stringent, this approach is acceptable. However, at higher frequencies, the differences in even- and odd-mode velocity become important, and their effects cannot be neglected. There are methods whereby these differences are incorporated into the design, on the one hand, or compensated for (by overlays, wiggly lines, *et cetera*), on the other hand.

The second important property of microstrips is, in effect, a disadvantage. A planar structure, it does not have an inherently high coupling, and we must resort to various ways of increasing the coupling, mainly in the form of interdigitated lines. The synthesis equations used in this section are not the most accurate, but they are quick and simple to use.

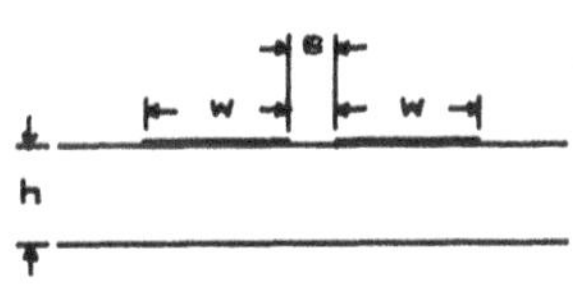

Figure B10 Cross section of coupled microstrip.

B6.1 Equations

The simplest way of calculating the properties of coupled microstrip lines was first proposed by Akhtarzad et al. [6], whereby the so-called shape ratios of uncoupled lines are used to obtain dimensions for coupled-line sections. If the desired even- and odd-mode impedances of a pair of coupled lines are given by Z_{0e} and Z_{0o}, respectively, then, using Wheeler's equation for both narrow and wide strips [7], the shape ratios $w/h_{so,se}$ are calculated from

$$\left.\frac{w}{h}\right|_{so,se} = \frac{8}{P} \cdot \left[\frac{(7 + 4/\epsilon_r)P}{11.0} + \frac{1 + 1/\epsilon_r}{0.81}\right]^{1/2} \tag{B.57}$$

where

$$P = \exp\left[\frac{Z_{0o,0e}}{84.8}(\epsilon_r + 1)^{1/2}\right] - 1 \tag{B.58}$$

In the equation for single strips of characteristic impedance Z_0, the factor P is given by

$$P = \exp\left(\frac{Z_0}{84.8}(\epsilon_r + 1)^{1/2}\right) - 1 \tag{B.59}$$

Alternatively, analysis equations given by Hammerstad [8] give the characteristic impedance of a single uncoupled line as

$$Z_0 = \begin{cases} \dfrac{60}{\sqrt{\epsilon_{re}}} \ln\left[\dfrac{8h}{w} + \dfrac{w}{4h}\right] & w/h \leq 1 \\ \dfrac{120\pi}{\sqrt{\epsilon_{re}}}\left[\dfrac{w}{h} + 1.393 + 0.677 \ln\left(\dfrac{w}{h} + 1.444\right)\right]^{-1} & w/h \geq 1 \end{cases} \tag{B.60}$$

and the effective dielectric constant as

$$\epsilon_{re} = \tfrac{1}{2}[\epsilon_r + 1) + (\epsilon_r - 1)F(w/h)] \tag{B.61}$$

where

$$Z_0 = \begin{cases} (1 + 12h/w)^{-1/2} + 0.04(1 - w/h)^2 & w/h \leq 1 \\ (1 + 12h/w)^{-1/2} & w/h \geq 1 \end{cases} \tag{B.62}$$

The equations by Osmani [9] combines those of Akhtarzad, so that they must be solved for G from

$$\left.\frac{w}{h}\right|_{so} = \frac{2}{\pi}\cosh^{-1}\left[\frac{(G+1)F-2}{(G-1)}\right]$$

$$+ r\cosh^{-1}\left\{\frac{\cosh^{-1}\frac{1}{2}[(G+1)F+(G-1)]}{\cosh^{-1}G}\right\} \tag{B.63}$$

where

$$F = \cosh\tfrac{1}{2}\pi(w/h)_{se} \tag{B.64}$$

$$r = 1/\pi \qquad \epsilon_r \geq 6 \tag{B.65}$$

$$r = \frac{8}{\pi(\epsilon_r + 1)} \tag{B.66}$$

The values of $(w/h)_{so}$ and $(w/h)_{se}$ are known through the even- and odd-mode impedances, and the value of G that satisfies (B.63) is obtained numerically. The dimensions of the coupled strip are then calculated from,

$$\frac{w}{h} = \frac{1}{\pi}(\cosh^{-1}H - \cosh^{-1}G) \tag{B.67}$$

$$\frac{s}{h} = \frac{2}{\pi}(\cosh^{-1}G) \tag{B.68}$$

where

$$H = \tfrac{1}{2}(G+1)F + \tfrac{1}{2}(G-1) \tag{B.69}$$

B6.2 Design Data

The equations in B6.1 were incorporated into LDIMS, and the following is an example of the use of the program. The relevant computer screen display is shown in Figure B11.

Example

Characteristic impedance,	$Z_0 = 50\ \Omega$
Even-mode impedance,	$Z_{0e} = 64.63\ \Omega$

```
RUN
DO YOU WISH TO DESIGN
        (1) A Coupled Coplanar Stripline
        (2) An Offset Parallel-Coupled Pair
        (3) A Coupled Microstripline
        (4) A Symmetrical Stripline
        (5) An Unsymmetrical Stripline
        (6) A Microstripline

Type the number desired, followed by RETURN3

**************************************************
           COUPLED MICROSTRIPLINE
**************************************************
               Give the Dielectric constant Er=2.32
         and the dielectric thickness in mm, b=10

These parameters are retained.

            Input the even mode impedance, Zoe=64.63
               and the odd mode impedance, Zoo=38.68

      DIMENSIONS:

                  Spacing, s = .2882503
                Linewidth, w = 4.172961

          Input the even mode impedance, Zoe=
```

Figure B11 **Computer screen display for calculating the dimensions of a pair of coupled microstrip lines.**

Odd-mode impedance,	$Z_{0o} = 38.68\ \Omega$
Dielectric thickness,	$h = 1.59$ mm
Dielectric constant,	$\epsilon_r = 2.32$

B6.3 Limits on Coupling

For soft substrates with the same restrictions on line widths and gaps as striplines, the maximum realizable coupling in microstrip is again in the order of -12 dB. For an alumina substrate with $\epsilon_r = 9.6$, a -10 dB coupling can be realized with two coupled lines. However, couplings as tight as -3 dB and more can be realized with Lange couplers, as explained in Section 4.4.1.

B7 SINGLE LINES

Three subroutines for calculating the widths of single, uncoupled lines have been included for convenience sake. They are

1. both thick and thin striplines, symmetrically arranged with respect to the center line;
2. unsymmetrical striplines for use with the offset parallel-coupled striplines; and
3. microstrips.

The operation of these subroutines is so simple that examples are not included.

```
10 REM ******************************************
20 REM *                    LDIMS                *
30 REM ******************************************
40 REM
50 REM This program combines three programs for the
   calculation
60 REM of stripline and microstrip dimensions from known
70 REM parameters and chosen material.
80 REM
90 REM Version # 3.3                    Date 08-12-1987
100 PI=3.141593
110 CLS
120 PRINT "                                 ***********"
130 PRINT "Cross sectional design program  *  LDIMS  *"
140 PRINT "                                 ***********"
150 PRINT " "
160 PRINT "********************************************"
170 PRINT "J.A.G. Malherbe, University of Pretoria, 1987"
180 PRINT "********************************************"
190 PRINT " "
200 REM CHOICE FOLLOWS.
210 PRINT "DO YOU WISH TO DESIGN"
220 PRINT "          (1) A Coupled Coplanar Stripline"
230 PRINT "          (2) An Offset Parallel-Coupled Pair"
240 PRINT "          (3) A Coupled Microstripline"
250 PRINT "          (4) A Symmetrical Stripline"
260 PRINT "          (5) An Unsymmetrical Stripline"
270 PRINT "          (6) A Microstripline"
280 PRINT " "
290 INPUT "Type the number desired, followed by
    RETURN",POINTER
300 PRINT " "
310 IF POINTER=1 THEN GOTO 340 ELSE IF POINTER=2 THEN
    GOTO 1220
320 IF POINTER=3 THEN GOTO 1780 ELSE IF POINTER=4 THEN
    GOTO 2190
330 IF POINTER=5 THEN GOTO 2410 ELSE IF POINTER=6 THEN
    GOTO 2560
340 REM ******************************************
350 REM *                   P-BARS                *
360 REM ******************************************
370 REM
380 REM Subroutine calculates the spacing for a given set
390 REM of capacitances.
400 REM Equations of Perlow IEEE Trans. MTT-35, 1987.
410 PI=3.14159
420 PRINT "**********************************************"

430 PRINT "              COUPLED COPLANAR STRIPLINE"
```

```
440 PRINT "*************************************************"
450 INPUT "                  Give the dielectric constant
    Er=",EPS
460 INPUT "                    the plate separation in mm, b=",B
470 INPUT "                        and conductor thickness t=",T
480 PRINT " "
490 PRINT "These parameters are retained"
500 B=B/1000:T=T/1000
510 PRINT " "
520 INPUT "              Input the even mode impedance, Zoe=",ZOE
530 INPUT "                 and the odd mode impedance, Zoo=",ZOO
540 COE=377/SQR(EPS)/ZOE:COO=377/SQR(EPS)/ZOO:DC=(COO-COE)/2
550 REM
560 INKR=.05*B
570 S=.3*B:GOSUB 680
580 DIFF=CEEX-DC
590 IF DIFF>0 THEN GOTO 610
600 IF DIFF<0 THEN INKR=-INKR
610 KYK=CEEX-DC:S=S+INKR
620 IF S<.3*B THEN GOSUB 760
630 IF S>.3*B THEN GOSUB 680
640 IF S=.3*B THEN GOSUB 680
650 IF ABS(INKR)<.001*B THEN GOTO 840
660 DOP=(CEEX-DC)*KYK:IF DOP>0 THEN GOTO 610
670 INKR=-INKR/2:GOTO 610
680 REM ****************************************
690 REM *          DELTA-C, S/B > 0.3          *
700 REM ****************************************
710 VAR1=PI/2*S/B:VAR2=EXP(VAR1)
720 CAPDUN=2/PI*LOG((VAR2^2+1)/(VAR2^2-1))
730 KORR=1+3.89531*T/B*(S/B)^(-.11467)
740 CEEX=CAPDUN*KORR
750 RETURN
760 REM ****************************************
770 REM *          DELTA-C, S/B <0.3           *
780 REM ****************************************
790 VAR1=PI/2*S/B:VAR2=EXP(VAR1)
800 CAPDUN=2/PI*LOG((VAR2^2+1)/(VAR2^2-1))
810 KORR=1+2.13507*T/B*(S/B)^(-.57518)
820 CEEX=CAPDUN*KORR
830 RETURN
840 PRINT " ":REM 1 capacitance Delta-C ="DC
850 PRINT "        DIMENSIONS:":PRINT " "
860 PRINT "                         Spacing, s ="S*1000
870 REM *****************************************
880 REM *   P-BARS EVEN FRINGING CAPACITANCE    *
890 REM *****************************************
900 IF S/B<.3 THEN GOTO 920
910 KORR=1+3.89531*T/B*(S/B)^(-.11467):GOTO 930
920 KORR=1+2.13507*T/B*(S/B)^(-.57518)
930 IF T/B>.5636 THEN GOTO 970
940 IF S/B<.08 THEN GOTO 960
```

```
950 KOMP=-.56*T/B:GOTO 1000
960 KOMP=-(.2933+3.333*S/B)*T/B:GOTO 1000
970 IF S/B<.08 THEN GOTO 990
980 KOMP=-.62+.54*T/B:GOTO 1000
990 KOMP=-.1653-5.6814*S/B+6.7475*S/B*T/B:GOTO 1000
1000 CFE=KORR*(S/B-LOG((VAR2+1/VAR2)/2)/PI*2)+KOMP
1010 KSI=1/(1-T/B):CF=(2*KSI*LOG(KSI+1)-(KSI-1)*
     LOG(KSI^2-1))/PI
1020 CPLATE=COE/2-CF-CFE:WEFF=CPLATE*(B-T)/2
1030 REM
1040 REM START OF ITERATION FOR PLATE WIDTH
1050 REM
1060 IF WEFF>.015*B^2/T THEN GOTO 1070,ELSE IF WEFF<.4*B
     THEN GOTO 1100
1070 W=WEFF
1080 PRINT "                    Linewidth, w ="W*1000
1090 GOTO 510
1100 W=.015*B^2/T:GOSUB 1160:INKR=.001*W/B
1110 VERSK=W-WCMP-WEFF
1120 W=W-INKR:GOSUB 1160:DOP=VERSK*(W-WCMP-WEFF)
1130 IF ABS(INKR)<.00001 THEN GOTO 1080
1140 IF DOP<0 THEN INKR=-INKR/2
1150 GOTO 1110
1160 REM ***************************************
1170 REM *               W-COMP                *
1180 REM ***************************************
1190 VAR4=EXP(PI/4*W/B):
     VAR5=LOG((VAR4^2+1)/(VAR4^2-1)*2)*2/PI
1200 VAR6=.44127+W/B-1/VAR5:
     WCMP=VAR6*(1-SQR(W/B*T/B/.015))^2*B
1210 RETURN
1220 REM ****************************************
1230 REM *                 OFFSET-P             *
1240 REM ****************************************
1250 REM
1260 REM Program uses the equations of J.P. Shelton to
1270 REM calculate the dimensions of overcoupled or offset
1280 REM parallel coupled lines. IEEE Trans. MTT-   1966.
1290 REM
1300 REM Version  #1.0                     Date 30-09-1987.
1310 REM
1320 PI=3.14159
1330 PRINT "**********************************************
     **"
1340 PRINT "        OFFSET PARALLEL-COUPLED STRIPLINE"
1350 PRINT "**********************************************
     **"
1360 INPUT "                  Give the dielectric constant
     Er=",EPS
1370 INPUT "                    the plate separation in mm,
     b=",BB
1380 INPUT "                    and the center strip ratio,
```

```
     s=",S
1390 PRINT " "
1400 PRINT "These parameters are retained."
1410 PRINT " "
1420 INPUT "                Input the even mode impedance,
     Zoe=",ZOE
1430 INPUT "                    and the odd mode impedance,
     Zoo=",ZOO
1440 PRINT " "
1450 ZO=SQR(ZOE*ZOO)
1460 RO=ZOE/ZOO
1470 REM
1480 A=EXP(60*PI^2/ZO/SQR(EPS)*(1-RO*S)/SQR(RO))
1490 B=(A-2+SQR(A^2-4*A))/2
1500 P=((B-1)*(1+S)/2+SQR((1+S)^2/4*(B-1)^2+4*S*B))/2
1510 R=S*B/P
1520 CFO=(-2/(1-S)*LOG(S)+
     1/S*LOG(P*R/(P+S)/(1+P)/(R-S)/(1-R)))/PI
1530 CO=120*PI*SQR(RO)/ZO/SQR(EPS)
1540 W=S*(1-S)*(CO-CFO)/2
1550 WO=((1+S)*LOG(P/R)+
     (1-S)*LOG((1+P)/(S+P)*(R-S)/(1-R)))/2/PI
1560 IF (W-WO)/S<.7 THEN 1630
1570 PRINT "        DIMENSIONS: "
1580 PRINT " "
1590 PRINT "                 Coupled width, Wc =",(W-WO)*BB
1600 PRINT "                      Line width, W =",W*BB
1610 PRINT "(Tight coupling)"
1620 GOTO 1410
1630 REM START OF LOOSE COUPLING APPROXIMATION.
1640 DC=CO*(1-1/RO)
1650 K=1/(EXP(PI/2*DC)-1)
1660 A=SQR((S-K)^2/(S+1)^2+K)-(S-K)/(S+1)
1670 Q=K/A
1680 WC=(S*LOG(Q/A)+(1-S)*LOG((1-Q)/(1+A)))/PI
1690 CFO=(1/(1+S)*LOG((1+A)/A/(1-Q))-1/(1-S)*LOG(Q))*2/PI
1700 CFA=-(1/(1+S)*LOG((1-S)/2)+1/(1-S)*LOG((1+S)/2))*2/PI
1710 W=(1-S^2)/4*(CO-CFO-CFA)
1720 PRINT "        DIMENSIONS: "
1730 PRINT " "
1740 PRINT "                 Coupled width, Wc =",WC*BB
1750 PRINT "                      Line width, W =",W*BB
1760 PRINT "(Loose Coupling)"
1770 GOTO 1410
1780 REM ****************************************
1790 REM *             COUPLED MICROSTRIP       *
1800 REM ****************************************
1810 PRINT "*********************************************
     **"
1820 PRINT "              COUPLED MICROSTRIPLINE"
1830 PRINT "*********************************************
     **"
```

```
1840 INPUT "                   Give the dielectric constant
     Er=",EPS
1850 INPUT "             and the dielectric thickness in mm,
      h=",HH
1860 PRINT " "
1870 PRINT "These parameters are retained."
1880 PRINT " "
1890 INPUT "                Input the even mode impedance,
     Zoe=",ZOE
1900 INPUT "                   and the odd mode impedance,
     Zoo=",ZOO
1910 PRINT " "
1920 ZO=ZOE/2:GOSUB 2690:WHSE=WH:ZO=ZOO/2:GOSUB 2690:WHSO=WH
1930 REM SHAPE RATIO FOR EVEN MODE WHSE
1940 REM SHAPE RATIO FOR ODD MODE WHSO
1950 REM CALCULATION OF COUPLED SHAPE RATIO NOW FOLLOWS
1960 DEF FNCO(X)=(EXP(PI/2*X)+EXP(-PI/2*X))/2
1970 DEF FNIC(X)=LOG(X+SQR(X^2-1))
1980 IF EPS>5.99999 THEN R=1/PI ELSE R=8/PI/(EPS+2)
1990 FFF=FNCO(WHSE)
2000 REM OPT BEGINS
2010 EKS=1.00001:DEL=EKS*.00001
2020 G=EKS:GOSUB 2060:DA=DIFF:G=EKS+DEL:GOSUB 2060:DB=DIFF
2030 EKSN=EKS-DEL*DA/(DB-DA)
2040 G=EKSN:GOSUB 2060:IF ABS(DIFF)<.00001 THEN GOTO 2130
2050 DEL=EKSN-EKS:EKS=EKSN:GOTO 2020
2060 REM ****************************************
2070 REM *          AKHTARZAD/OSMANI            *
2080 REM ****************************************
2090 NICA=FNIC((G+1)/2*FFF+(G-1)/2)/FNIC(G)
2100 DIFF=WHSO-2/PI*FNIC(((G+1)*FFF-2)/(G-1))-R*FNIC(NICA)
2110 RETURN
2120 REM ****************************************
2130 REM SHAPE FACTOR CALCULATED.
2140 HHH=(G+1)*FFF/2+(G-1)/2:SH=2/PI*FNIC(G):
     WH=(FNIC(HHH)-FNIC(G))/PI
2150 PRINT "       DIMENSIONS:":PRINT " "
2160 PRINT "                        Spacing, s ="SH*HH
2170 PRINT "                      Linewidth, w ="WH*HH
2180 GOTO 1890
2190 REM ****************************************
2200 REM *       SYMMETRICAL STRIPLINE          *
2210 REM ****************************************
2220 PRINT "************************************************
     **"
2230 PRINT "                SYMMETRICAL STRIPLINE"
2240 PRINT "************************************************
     **"
2250 INPUT "                  Give the dielectric constant,
     Er=",EPS
2260 INPUT "                     the plate separation in mm, b=",B
2270 INPUT "                         and conductor thickness t=",T
2280 PRINT " "
2290 PRINT "These parameters are retained"
2300 B=B/1000:T=T/1000
2310 PRINT " "
2320 INPUT "       Input the characteristic impedance, Zo=",ZO
```

```
2330 REM
2340 CTOT=377/SQR(EPS)/ZO:TEBE=1/(1-T/B)
2350 CFR=(2*TEBE*LOG(TEBE+1)-(TEBE-1)*LOG(TEBE^2-1))/PI
2360 W=B*(1-T/B)*(CTOT-4*CFR)/4
2370 IF W/B/(1-T/B)<.35 THEN 2400
2380 PRINT "                Linewidth, w ="W*1000
2390 GOTO 2310
2400 W=B*(.07*(1-T/B)+W/B)/1.2:GOTO 2380
2410 PRINT "************************************************
     **"
2420 PRINT "                OFFSET STRIPLINE"
2430 PRINT "************************************************
     **"
2440 INPUT "               Give the dielectric constant
     Er=",EPS
2450 INPUT "                 the plate separation in mm,
     b=",BB
2460 INPUT "                 and the center strip ratio,
     s=",S
2470 PRINT " "
2480 PRINT "These parameters are retained."
2490 PRINT " "
2500 INPUT "        Input the characteristic impedance,
     Zo=",ZO
2510 CTOT=377/SQR(EPS)/ZO:CFR=.44127
2520 CS=S*BB
2530 W=(CTOT-4*CFR)/(1/(BB-CS)+1/(BB+CS))/2
2540 PRINT "                Linewidth, w ="W
2550 GOTO 2490
2560 REM ****************************************
2570 REM *             MICROSTRIP              *
2580 REM ****************************************
2590 PRINT "***********************************************
     **"
2600 PRINT "                 MICROSTRIPLINE"
2610 PRINT "***********************************************
     **"
2620 INPUT "               Give the dielectric constant
     Er=",EPS
2630 INPUT "         and the dielectric thickness in mm,
     h=",HH
2640 PRINT " "
2650 PRINT "These parameters are retained."
2660 PRINT " "
2670 INPUT "        Input the characteristic impedance,
     Zo=",ZO
2680 GOSUB 2690:W=WH*HH:GOTO 2750
2690 REM ****************************************
2700 REM *  SUBROUTINE SINGLE LINE MICROSTRIP   *
2710 REM ****************************************
2720 PARP=EXP(ZO/42.4*SQR(EPS+1))-1
2730 WH=8/PARP*SQR(PARP*(7+4/EPS)/11+(1+1/EPS)/.81)
2740 RETURN
2750 PRINT " "
2760 PRINT "                Linewidth, w ="W
2770 PRINT " "
2780 GOTO 2670
```

REFERENCES

[1] J. A. G. Malherbe, *Microwave Transmission Line Filters,* Artech House, Dedham, MA, 1979.

[2] S. B. Cohn, "Shielded Coupled-Strip Transmission Line," *IRE Trans. Microwave Theory Tech.,* Vol. MTT-3, October 1955, pp. 29–38.

[3] J. P. Shelton, Jr., "Impedances of Offset Parallel-Coupled Strip Transmission Lines," *IEEE Trans. Microwave Theory Tech.,* Vol. MTT-14, January 1966, pp. 7–15.

[4] W. J. Getsinger, "Coupled Rectangular Bars between Parallel Plates," *IRE Trans. Microwave Theory Tech.,* Vol. MTT-10, January 1962, pp. 65–72.

[5] S. M. Perlow, "Analysis of Edge-Coupled Shielded Strip and Slab-line Structures," *IEEE Trans. Microwave Theory Tech.,* Vol. MTT-17, May 1969, pp. 522–529.

[6] S. Akhtarzad, et al., "The Design of Coupled Microstrip Lines," *IEEE Trans. Microwave Theory Tech.,* Vol. MTT-23, June 1975, pp. 486–492.

[7] H. A. Wheeler, "Transmission Line Properties of a Strip on a Dielectric Sheet on a Plane," *IEEE Trans. Microwave Theory Tech.,* Vol. MTT-25, August 1977, pp. 631–647.

[8] E. O. Hammerstad, "Equations for Microstrip Circuit Design," *Proc. 5th Eur. Microwave Conf.,* 1975, pp. 268–272.

[9] R. M. Osmani, "Synthesis of Lange Couplers," *IEEE Trans. Microwave Theory Tech.,* Vol. MTT-29, February 1981, pp. 168–170.

INDEX

www.ingramcontent.com/pod-product-compliance
Lightning Source LLC
LaVergne TN
LVHW101501200926
848483LV00001B/194

* 9 7 8 0 8 9 0 0 6 3 0 0 2 *